Higher Grade Biology

Higher Grade
BIOLOGY

Team Co-Ordinator
James Torrance

Writing Team
James Torrance
James Fullarton
Clare Marsh
James Simms
Caroline Stevenson

Diagrams by James Torrance

Hodder & Stoughton
LONDON SYDNEY AUCKLAND TORONTO

British Library Cataloguing in Publication Data

Torrance, James
  Higher grade biology
  1. Biology
  I. Title
  574

  ISBN 0 340 53611 X

First published 1991
Sixth impression 1992

Phototypeset by Input Typesetting Ltd, London
Printed in Great Britain for the educational publishing
division of Hodder and Stoughton Ltd,
Mill Road, Dunton Green, Sevenoaks, Kent by
Thomson Litho Ltd.

Contents

Preface

This book has been written to articulate closely with Standard Grade Biology. It is intended to act as a valuable resource for pupils continuing their studies to Higher Grade. It provides a concise set of notes which adhere to the syllabus for SCE Higher Grade Biology to be examined in and after 1991.

Each section of the book matches a syllabus topic. Each chapter corresponds to part of a syllabus sub-topic and is followed by a selection of questions designed to consolidate *knowledge* and *understanding* and give practice in *problem solving*.

In addition, the book contains many examples of *data interpretation* and *experimental design* questions to meet the demands of the new examination. Summaries of key facts and concepts are given at regular intervals throughout the book to reinforce learning and several appendices have been included to act as reference sections.

Also available to support this book: *Higher Grade Biology Multiple Choice Tests* ISBN 0340 550597, and accompanying free answers sheet: ISBN 0340 559462.

Section 1 Cell Biology

1 Cell variety in relation to function

At first glance a random selection of structures from living things, such as a chicken bone, a human nerve, a frog kidney, an octopus eye, a rose petal, a moss leaf and a dandelion root, have little in common with respect to their outward appearance. However microscopic examination reveals that they all share one essential feature: they are made up of **cells**.

The cell is the basic unit of life. It is the smallest structure that is able to lead an **independent life** and show all the **characteristics** of living things.

A cell is like a chemical factory. It possesses a controlling centre, the **nucleus**, which governs the cell 'machinery'. The latter consists of a variety of subcellular structures which are illustrated in Appendix I. Their functions will be discussed in later chapters.

Unicellular organisms

Amongst living things there exists a variety of tiny organisms, each of which consists of only one cell. To survive, such an organism must be capable of manufacturing all the necessary chemicals (**metabolites**) and performing all the functions essential for the continuation of life, *within one cell*.

This is made possible by the presence of a variety of structures each of which performs a specific role. Such **unicellular level** of **organisation** is illustrated by *Pleurococcus* (see figure 1.1) which is a tiny plant and *Paramecium* (see figure 1.2) which is a tiny animal.

Some unicellular organisms (see question 7 on page 5) cannot be so easily classified since they possess plant and animal-like characteristics.

Multicellular organisms

Advanced plants and animals consist of an enormous number (often millions) of cells. It would be inefficient for every one of these cells to perform every function essential for the maintenance of life. Instead the cells are arranged into **tissues**.

A tissue is a group of cells specialised to

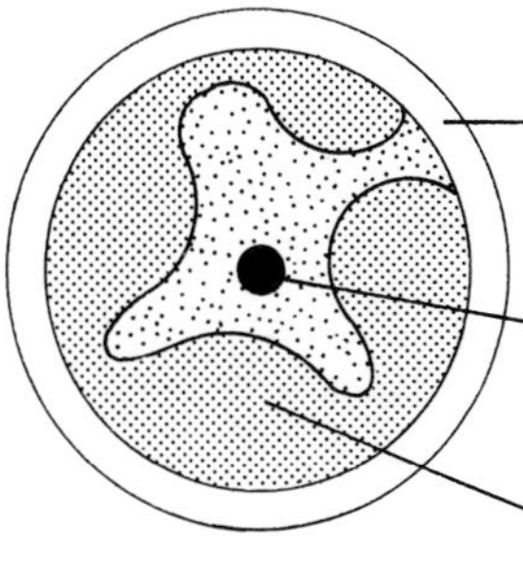

structure	function	details
cell wall	support and protection	composed of cellulose fibres which combine strength with a degree of elasticity
nucleus	control of cell activities	contains genetic material which controls day to day running of cell and is passed on to daughter cells formed by asexual reproduction
chloroplast	photosynthesis	a massive lobed structure which produces all of the cell's carbohydrate food

Figure 1.1   Unicellular level of organisation in Pleurococcus

1

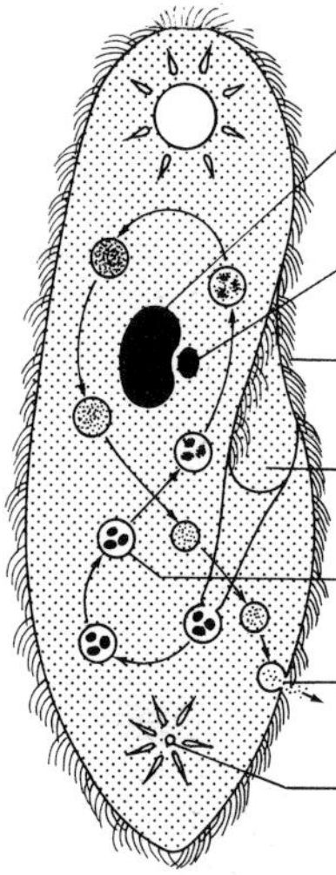

structure	function	details
macronucleus	day to day control of cell	contains genetic material which controls normal cell metabolism
micronucleus	long term control	contains genetic material passed on to next generation at reproduction
cilia	locomotion	hair-like structures connected at bases by threads allowing co-ordinated beating
ciliated oral groove and pharynx	ingestion of food	food transported to region of food vacuole formation
food vacuole	transport and digestion of food	food digested and end products distributed as food vacuole is moved round cell
anal pore	waste removal	undigested material is expelled
contractile vacuole	osmoregulation	excess water gathered and expelled to exterior of cell

Figure 1.2   *Unicellular level of organisation in* Paramecium

perform a particular function (or functions). This **multicellular level of organisation** is therefore said to show a **division of labour**.

Figures 1.3 and 1.4 illustrate a selection of tissues (and their component cells) from two multicellular organisms.

Variation between cells

Within one tissue

Although the cells that make up a tissue may all be of the same type, smooth muscle for example consists of spindle-shaped cells only, a tissue often possesses a *variety* of cell types. Blood contains red blood cells, white blood cells and platelets. Phloem tissue consists of sieve tubes and companion cells.

Between different tissues

When one tissue is compared with another, for example blood compared with ciliated epithelium, or phloem compared with xylem, then *variation* in cell structure becomes even more apparent.

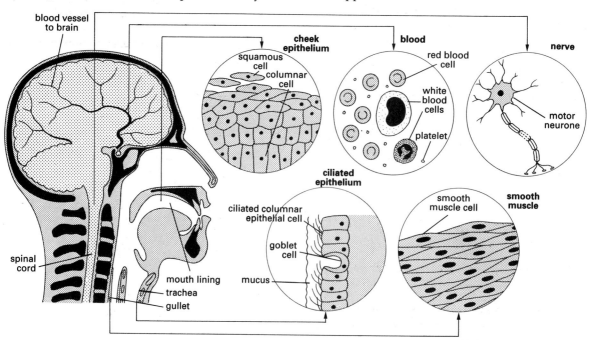

Figure 1.3   *Human tissues and cells (not drawn to scale)*

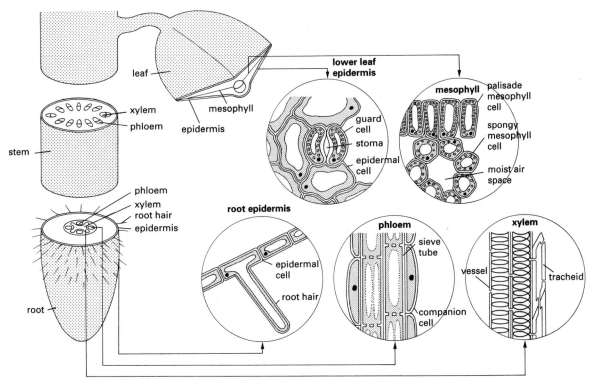

Figure 1.4   Plant tissues and cells (not drawn to scale)

Structure in relation to function

Variation in relation to function exists because each cell's structure is exactly tailored to suit its function. Once a cell has finished dividing and growing, it becomes specialised by undergoing physical and chemical changes until its size, shape and chemical 'machinery' are perfectly suited to its future role in a particular tissue.

Tables 1.1 and 1.2 explain how the structure of each of the cell types shown in figures 1.3 and 1.4 is related to the cell's function.

With each specialised tissue performing its role so effectively, a multicellular organism is able to function at a much more advanced level than a unicellular organism.

QUESTIONS

1 *'A red blood corpuscle is an example of a cell whose structure can easily be related to its function.'* Justify this statement.
2 **a)** State TWO functions of xylem tissue.
  b) Describe how a xylem vessel is structurally suited to perform each of these functions.
3 **a)** Make a simple diagram of a small sample of the tissue that covers the inner surface of the human windpipe, as it would be seen in longitudinal section under the microscope.
  b) Name both types of cell that you have drawn and describe how each is ideally suited to the function that it performs.
4 Figure 1.5 shows a transverse section of a young root.

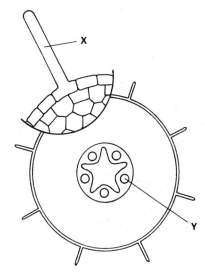

Figure 1.5

tissue	cell type	specialised structural features	function
cheek epithelium	epithelial cell	flat irregular shape (allowing cells to form loose covering layer, constantly replaced from below during wear and tear)	protection of mouth lining
blood	red blood cell	small size and biconcave shape present large surface area; rich supply of haemoglobin present	uptake and transport of oxygen to living cells
	white blood cell	able to change shape; sacs of microbe-digesting enzymes present in some types	destruction of invading pathogens (disease-causing micro-organisms)
nerve	motor neurone	long insulated tail-like extension of cytoplasm	transmission of nerve impulses
ciliated epithelium	goblet cell	cup shape; able to produce mucus	secretion of mucus which traps dirt and germs
	ciliated epithelial cell	hair-like cilia which beat upwards	sweeping of dirty mucus up away from lungs
smooth muscle	smooth muscle cell	spindle shape (allowing cells to form sheets capable of contraction)	movement of food down gullet by peristalsis

Table 1.1   Structure of animal cells in relation to function

tissue	cell type	specialised structural features	function
lower leaf epidermis	epidermal cell	irregular shape (allowing cells to fit like jigsaw into strong layer)	protection
	guard cell	sausage shape; thick inner cell wall facing stoma; chloroplasts present	control of gaseous exchange by changing shape, and opening or closing stomata
mesophyll	palisade mesophyll cell	chloroplasts present; columnar shape (allows densely packed green layer to be presented to light)	primary region of light absorption and photosynthesis
	spongy mesophyll cell	'round' shape allows loose arrangement in contact with moist air spaces for absorption of carbon dioxide	secondary region of photosynthesis
phloem	sieve tube	sieve plates and continuous system of cytoplasmic strands	transport (translocation) of soluble carbohydrates
	companion cell	large nucleus in relation to cell size	control of sieve tube
xylem	vessel	hollow tube; walls strengthened by lignin; lignin deposited as rings or spirals allowing expansion and contraction	mechanical support and water transport
	tracheid	hollow spindle-shaped cell whose ends overlap with neighbours; lignified walls	mechanical support and water transport
root epidermis	epidermal cell	box-like shape allowing cells to fit together like brick wall	protection
	root hair	long extension presenting large surface area in contact with soil solution	absorption of water and mineral salts

Table 1.2   Structure of plant cells in relation to function

a) Identify structure **X** and state its function.

b) Name TWO types of cell that would be present in tissue **Y** and compare them with respect to their structure and function.

c) Does a division of labour exist in a root? Explain your answer.

5 Figure 1.6 shows the epithelial tissue which makes up the surface layer of a villus in the human small intestine.

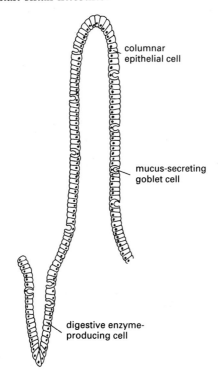

columnar epithelial cell

mucus-secreting goblet cell

digestive enzyme-producing cell

Figure 1.6

State TWO ways in which this tissue (i) resembles, (ii) differs from, the epithelium that lines the windpipe (see figure 1.3).

6 a) Identify the type of sex cell shown in figure 1.7.

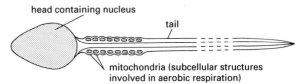

head containing nucleus

tail

mitochondria (subcellular structures involved in aerobic respiration)

Figure 1.7

b) Give TWO reasons why this cell is well suited to its function.

7 The unicellular organism shown in figure 1.8 is called *Euglena*.

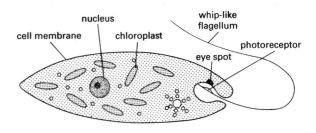

nucleus

whip-like flagellum

cell membrane

chloroplast

photoreceptor

eye spot

Figure 1.8

a) Suggest how it obtains its food.

b) Suggest how it (i) senses light, (ii) moves towards light.

c) Briefly explain why *Euglena* cannot be satisfactorily classified as either a plant or an animal.

2 Absorption and secretion of materials

Absorption means the uptake of materials by a cell from its external environment. **Secretion** means the discharge of useful intracellular molecules into the surrounding medium by a cell.

The movement of small molecules or ions (tiny electrically-charged particles) into or out of a cell normally occurs as a result of **diffusion, osmosis** or **active transport** depending upon the nature of the substance involved.

In addition, some cells can engulf large particles by a process called **phagocytosis** ('cell-eating', see page 53). When the engulfed material is liquid, the process is called **pinocytosis** ('cell-drinking', see figure A1.1 on page 245).

Many types of cell also synthesise large molecules of protein which they **secrete** out of the cell (see page 42).

Cell boundaries

All living cells are surrounded by a **cell membrane (plasma membrane)**. In addition, plant cells possess a **cell wall**. The structure of each of these boundaries is closely related to the role that it plays in the movement of materials into and out of the cell.

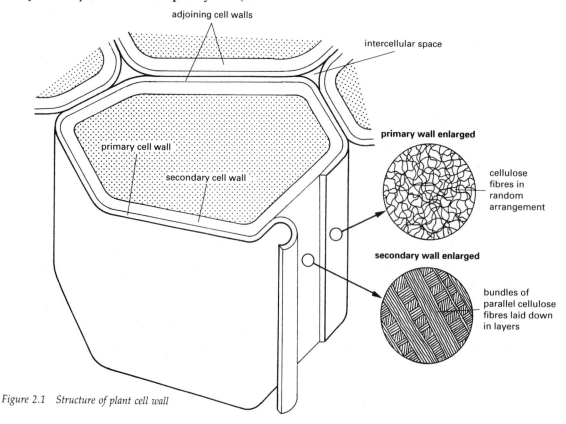

Figure 2.1 Structure of plant cell wall

Structure of the cell wall

The **cell wall** is a non-living layer composed mainly of **cellulose**. This complex carbohydrate consists of unbranched chains of glucose molecules grouped together as **fibres**.

The primary (first-formed) cell wall consists of cellulose fibres enmeshed at random amongst a variety of materials including pectic substances (e.g. calcium pectate), proteins (e.g. enzymes) and water-filled spaces.

The secondary cell wall is laid down inside the primary wall. It consists of **layers** of closely packed cellulose fibres running in different well defined directions (see figure 2.1) This arrangement makes the cell wall strong, fairly rigid and yet slightly elastic. It is able to stretch slightly when the cell absorbs water. However when it reaches the limit of its elasticity, the cell wall resists further uptake of water by the cell and prevents the latter from bursting (see also page 10).

Since cellulose is **hydrophilic** (has an affinity for water), the spaces between the fibres in a cell wall are normally water-filled. Adjoining cell walls and intercellular spaces (see figure 2.1) provide a **continuous water-conducting** pathway throughout the plant. This allows water to move easily from tissue to tissue without having to enter and leave every living cell along the way.

Structure of the plasma membrane

The **plasma membrane** is composed of **protein** and **phospholipid** (fat) molecules. Although the precise arrangement of these molecules is still unknown, most evidence supports the **fluid mosaic model** of cell membrane structure (see figure 2.2). This proposes that the plasma membrane consists of a **fluid layer** of constantly moving phospholipid molecules containing a patchy **mosaic** of protein molecules.

The protein molecules vary in size and structure. Some extend partly into the phospholipid layer while others extend across it from one side to the other. Some of these enclose narrow **channels** making the membrane **porous**. The fluid nature of the membrane allows some movement of proteins (and pores).

The protein molecules also vary in function. The channel-making proteins provide the means by which small molecules are able to pass through the membrane. Other protein molecules act as receptors for hormones or antibody reactions. Some serve as enzymes or structural proteins whilst others actively 'pump' certain molecules across the membrane (see page 12).

Investigating the chemical nature of the cell membrane
The cell sap present in the central vacuole of a beetroot cell (see figure 2.3) contains red

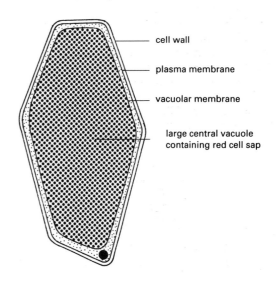

Figure 2.3   Beetroot cell

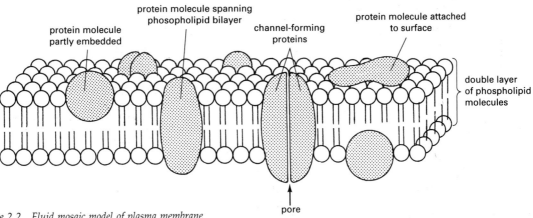

Figure 2.2   Fluid mosaic model of plasma membrane

pigment. 'Bleeding' (the escape of this red cell sap from a cell) indicates that the cell's plasma and vacuolar membranes have been damaged.

In the experiment shown in figure 2.4, four identical cylinders of fresh beetroot are prepared using a cork borer. The cylinders are thoroughly washed in distilled water to remove traces of red cell sap from outer damaged cells. The figure shows the results of subjecting the cylinders to various conditions.

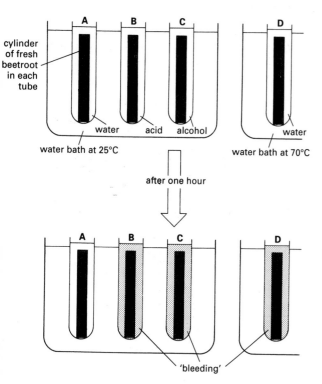

Figure 2.4 *Investigating the chemical nature of the cell membrane*

Since 'bleeding' occurs in **B, C** and **D**, it is concluded that the cell membranes have been destroyed in each of these tubes.

Acid and high temperature destroy the membrane by **denaturing** its **protein** molecules. This allows the red cell sap to leak out. Alcohol **dissolves** the **phospholipid** (fat) component of the membrane and therefore allows the pigment to escape.

Diffusion

Diffusion is the net movement of molecules or ions from a region of high concentration to a region of low concentration of that type of molecule or ion.

The difference that exists between two regions before diffusion occurs is called the **concentration gradient**. During diffusion, molecules and ions always move along a concentration gradient from high to low concentration.

Diffusion is a basic cell process. It is the means by which useful substances such as oxygen enter a cell, and waste materials such as carbon dioxide leave a cell.

Effect of cell wall and plasma membrane on diffusion

Many water-filled spaces and some large pores occur amongst the cellulose fibres in a cell wall. These make it **freely permeable** to all molecules in solution. The cell wall does not, therefore, act as a barrier to diffusion.

The plasma membrane possesses *very tiny* pores (see figure 2.2) and is freely permeable to small molecules such as oxygen, carbon dioxide and water. However it is *not* equally permeable to all substances. This is shown in figure 2.5.

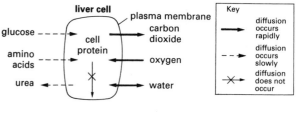

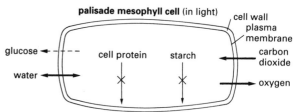

Figure 2.5 *Diffusion into and out of cells*

Larger molecules such as urea, amino acids and glucose diffuse through the membrane more slowly. Even larger molecules such as starch and protein are unable to pass through. The plasma membrane is said therefore to be **selectively permeable**.

Osmosis

Osmosis is the net movement of water molecules from a region of higher water concentration

(HWC) to a region of lower water concentration (LWC) through a selectively permeable membrane.

Investigating osmosis

In the experiment shown in figure 2.6, visking tubing sausage **A** is found to gain weight after one hour whereas **B** remains unchanged.

It is concluded that **A** has gained weight because small water molecules have passed rapidly through the tiny pores in the selectively permeable membrane along a concentration gradient from a region of higher water concentration (HWC) to a region of lower water

concentration (LWC).

The solution with the HWC is said to be **hypotonic** to the solution with the LWC which is said to be **hypertonic**.

It is concluded that sausage **B** shows no change in weight because both solutions have an equal water concentration and neither makes a net gain of water molecules. Two solutions of equal water concentration are said to be **isotonic**.

Osmosis and cells

Movement of water by osmosis occurs in living things between neighbouring cells, and between cells and their immediate extracellular

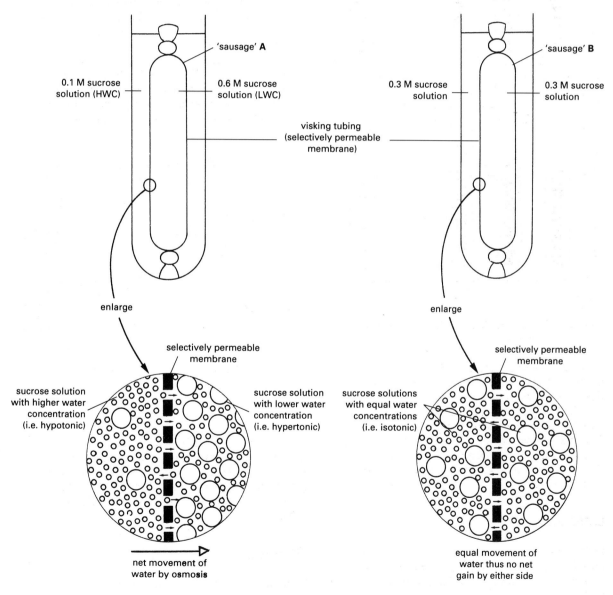

Figure 2.6   Investigating osmosis

environment. For example, in humans, water passes continuously by osmosis from the gut cells into the bloodstream. Table 2.1 lists several examples of osmosis in a green plant. Like diffusion, osmosis is a basic cell process.

direction of movement of water	significance
soil solution → root hairs	water absorbed by plant from soil
xylem vessels → stem cells	water makes cells turgid giving support
xylem vessels → green leaf cells	water used as raw material for photosynthesis

Table 2.1   Movement of water by osmosis in plant cells

Role of plasma membrane in osmosis

Whenever a cell is in contact with a solution (or another cell) of differing water concentration, osmosis occurs. This is made possible by the fact that the plasma membrane is selectively permeable. It allows the rapid movement of water molecules through it but only allows larger molecules to move across slowly or not at all.

The direction in which net movement of water molecules occurs depends upon the water concentration of the liquid in which the cell is immersed compared with that of the cell contents (see figures 2.7 and 2.8). The water relation of cells (and solutions) can also be described in terms of water potential (see Appendix 2).

Role of cell wall

Unlike an animal cell, a plant cell placed in water does not burst. As water enters a plant cell by osmosis, its central vacuole swells up and presses the cytoplasm and plasma membrane against the cell wall which stretches slightly. As this process continues, the wall *presses back* (i.e. exerts a **wall pressure**) on the cell contents. Eventually a point is reached where the wall pressure stops further water from entering and prevents the cell from bursting.

No wall pressure exists in a plasmolysed cell where the contents have shrunk and pulled away from the cell wall.

Measuring the water concentration of potato cell sap

Five thin cylinders are cut out of the same potato using a cork borer. The cylinders are trimmed to equal length and rolled dry on filter paper. Each is carefully weighed using an electronic balance and then immersed in a test tube containing one of the sucrose solutions listed in table 2.2.

After one hour each cylinder is rolled dry and reweighed. If possible these results are pooled with those of other groups carrying out the same investigation. The **ratio** of each cylinder's final weight to initial weight is then calculated as shown in table 2.2. which gives a specimen set of results. The results are then graphed and the best straight line drawn through the points (see figure 2.9).

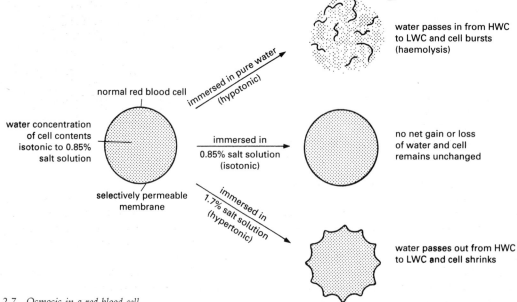

Figure 2.7   Osmosis in a red blood cell

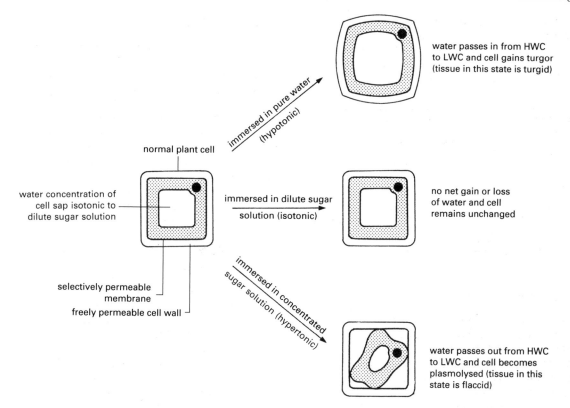

Figure 2.8 Osmosis in a plant cell

molar con-centration of sucrose solution (M)	initial weight of potato cylinder (g)	final weight of potato cylinder (g)	final weight ÷ initial weight
0.1	0.67	0.70	1.05
0.2	0.67	0.68	1.02
0.3	0.68	0.67	0.99
0.4	0.67	0.63	0.94
0.5	0.68	0.61	0.90

Table 2.2 Specimen set of results

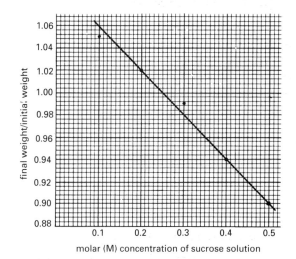

Figure 2.9 Graph of results

In those sucrose solutions where the ratio of final weight/initial weight > 1.0, it is concluded that water has *passed into* the potato cells by osmosis since the bathing sucrose solution has a higher water concentration than that of the cell sap.

In those sucrose solutions where the ratio of final weight/initial weight < 1.0, it is concluded that water has *passed out* of the cells by osmosis since the bathing sucrose solution has a lower water concentration than that of the cell sap.

A ratio of 1.0 indicates that *no net movement of water* into or out of the cell has occurred by osmosis because the water concentration of the cell sap is *equal* to that of the bathing sucrose solution. This concentration of sucrose can be easily read from the graph. Thus, in this case, the water concentration of the potato cell sap is equal to the water concentration of 0.25 molar sucrose solution.

Table 2.3 lists experimental design features and precautions that should be adopted during this investigation.

design feature or precaution	reason
potato cut into thin cylinders	to expose a large surface area of potato cells to bathing medium
class results pooled	to make results more *valid*
cylinders blotted before weighing	to remove excess liquid since amount adhering to cylinders will vary
all factors kept equal except concentration of sucrose	to ensure that investigation only involves one variable factor

Table 2.3   *Experimental design features and precautions*

Active transport

Active transport is the movement of molecules and ions across the plasma membrane from a low to a high concentration i.e. against a **concentration gradient**.

Active transport works in the *opposite* direction to the passive process of diffusion and requires **energy**.

Consider the two situations shown in figure 2.10. Ion type **A** is being actively transported *into* the cell whereas ion type **B** is being actively transported *out of* the cell.

Active transport in a plant cell

From the experiment shown in figure 2.11, it is concluded that the marine plant, *Valonia*, is able to **select** and **accumulate** potassium ions in its cell sap to a concentration greatly in excess of the concentration in the external environment. This **selective ion uptake** is brought about by active transport.

Active transport of sodium in the opposite direction accounts for the fact that the cell

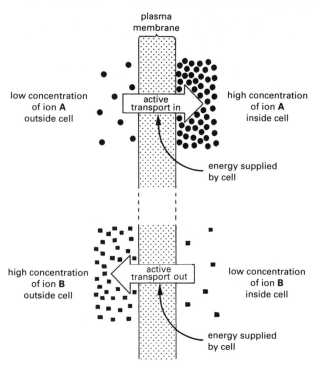

Figure 2.10   *Active transport of two different ions*

simultaneously maintains a low concentration of sodium ions in its cell sap despite the high concentration present in the surrounding sea water.

The plant constantly loses potassium ions and gains sodium ions by the passive process of diffusion. However, by employing active transport, it is able to maintain the optimum level of these ions in its cell sap by continuously 'pumping' potassium in and sodium out against their respective concentration gradients.

Role of the plasma membrane in active transport

Recent studies suggest that certain protein molecules present in the plasma membrane act

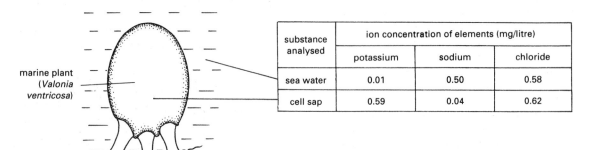

substance analysed	ion concentration of elements (mg/litre)		
	potassium	sodium	chloride
sea water	0.01	0.50	0.58
cell sap	0.59	0.04	0.62

marine plant (*Valonia ventricosa*)

Figure 2.11   *Investigating active transport in* Valonia

as **carrier molecules**. These protein molecules 'recognise' specific ions and transfer them across the plasma membrane (see figure 2.12). The energy required for this active process is supplied by **ATP** (adenosine triphosphate) formed during respiration (see page 17).

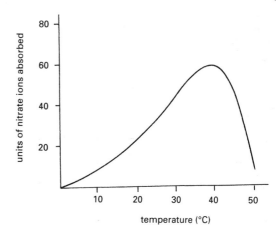

Figure 2.13   *Effect of temperature on ion uptake*

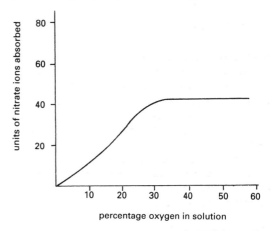

Figure 2.14   *Effect of oxygen concentration on ion uptake*

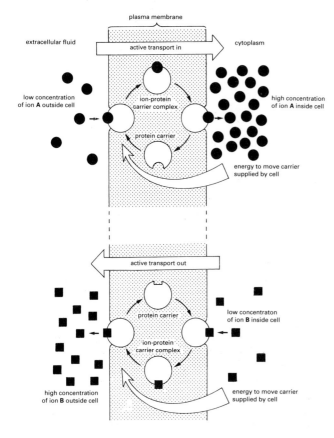

Figure 2.12   *Role of protein carriers in active transport*

Conditions required for active transport

Factors such as **temperature**, availability of **oxygen** and level of **respiratory substrate** (e.g. glucose), which directly affect a cell's respiration rate, also affect the rate of active transport.

Figures 2.13 and 2.14 show the effects of increasing temperature and oxygen concentration on the rate of active uptake of nitrate ions by the cells of barley roots. Increasing the temperature brings about an increase in ion uptake until at high temperatures enzymes become denatured and the cell dies. Increasing the oxygen concentration results in an increased rate of ion uptake until some other factor affecting the process becomes limiting.

QUESTIONS

1 Name THREE methods by which molecules may enter a cell.
2 a) Briefly describe the structure of a plant cell wall.
   b) Figure 2.15 shows a possible arrangement of the molecules in a plasma membrane.

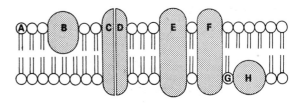

Figure 2.15

(i) What name is given to this model?
(ii) Identify molecule types **A** and **B**.
(iii) Which lettered structures enclose a narrow channel?
(iv) What is the function of such a channel?

3 a) Name a chemical molecule (other than water) that would be found diffusing out of a palisade leaf cell in (i) light, (ii) darkness.
b) Give an example of a chemical molecule that is too large to diffuse out of a palisade leaf cell.

4 Figure 2.16 shows the direction of movement of two different substances through the plasma membrane of an animal cell.
a) Name the process **X** and **Y**.
b) Which of these processes requires energy?
c) Which process will be unaffected by a decrease in oxygen concentration in the animal's environment?
d) Predict what will happen to the rate of process **Y** if the temperature of the cell is reduced to 4° C for several hours. Give a reason for your answer.

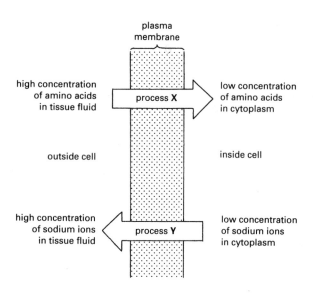

Figure 2.16

5 In an experiment, groups of potato discs of known masses were immersed in sucrose solutions of different concentration for a few hours and then reweighed. The results were plotted as shown in figure 2.17.
a) Was 0.2 M sucrose solution hypotonic or hypertonic to the potato cell sap? Give a reason for your answer.
b) To which molarity of sucrose was the potato cell sap equal in water concentration?

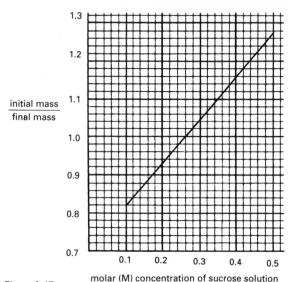

Figure 2.17

6 It is possible that some protein carrier molecules actively transport materials against a concentration gradient by rotating within the plasma membrane. Rearrange the six stages in figure 2.18 to give the correct sequence in which this would occur, (start with **A**).

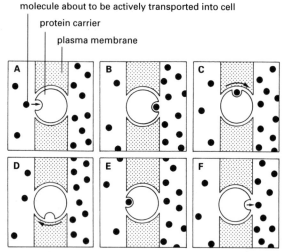

Figure 2.18

EXPERIMENTAL DESIGN QUESTION

Untreated cells from a fresh beetroot were viewed under the microscope and then cell samples (from the same beetroot) which had been immersed in one of three different concentrations of sucrose solution were examined.

Observations: Compared with untreated cells, those in 0.15 M (molar) sucrose appeared more turgid, those in 0.35 M sucrose appeared to be plasmolysed, but those in 0.25 M sucrose seemed to be unchanged.

Hypothesis: Beetroot cells gain water when immersed in 0.15 M sucrose solution, lose water in 0.35 M sucrose but neither gain nor lose water in 0.25 M sucrose.

Problem: Design an experimental procedure that would enable you to test the validity of the hypothesis. (See Appendix 3 for help.)

DATA INTERPRETATION

Table 2.4 refers to the concentrations of certain chemical ions in the cells of the alga, *Nitella*, and in the pond water in which this green plant lives.

In figure 2.19 graph 1 shows the effect of light intensity on ion uptake by *Nitella*.

Graph 2 presents the results of an experiment set up to investigate the effect of oxygen concentration on uptake of potassium ions and consumption of sugar by the cells of excised (cut off) roots of barley seedlings.

Graph 3 gives the results from an experiment

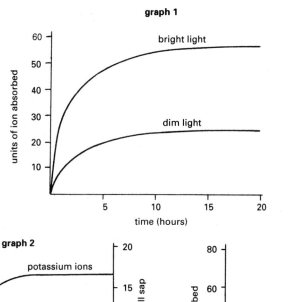

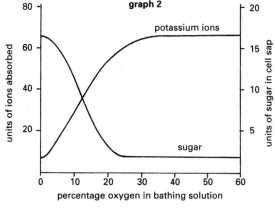

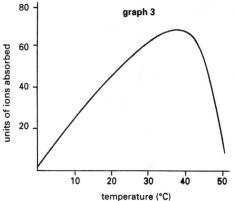

Figure 2.19

		ion concentration of element (mg/litre)				
		calcium (Ca⁺⁺)	chloride (Cl⁻)	magnesium (Mg⁺⁺)	potassium (K⁺)	sodium (Na⁺)
substance	cell sap	380	3750	260	2400	1988
analysed	pond water	26	35	36	2	28
	accumulation ratio	14.6	107.1	7.2		

Table 2.4

set up to investigate the effect of temperature on potassium ion uptake by barley roots.

a) (i) From the table, make a generalisation about the concentration of ions in the pond water compared with their concentration in the cell sap. (1)

(ii) Calculate the accumulation ratio for potassium and sodium. (1)

(iii) In what way does the data support the theory that a cell membrane is *selective* with respect to the process of ion uptake? (1)

(iv) Does the data support or dispute the suggestion that ion uptake occurs as a result of diffusion? Explain your answer. (1)

b) Account for the effect of light intensity on ion uptake by *Nitella* in graph 1. (1)

c) (i) From graph 2, state the effect that an increase in oxygen concentration from 0 to 30% has on the rate of ion uptake. (1)

(ii) Suggest why ion uptake levels off beyond 30% oxygen. (1)

(iii) What relationship exists between units of ion absorbed and units of sugar present in cell sap? Suggest why. (1)

d) (i) State the temperature at which greatest uptake of ions occurs in graph 3.

(ii) Account for the sudden decline in ion uptake shown by the graph. (1)

e) A farmer decided to drain a field which had become waterlogged each year following heavy rainfall. With reference to ion uptake, give TWO possible reasons why his crop of barley showed a greatly increased yield the next year. (1)

(10)

What you should know (CHAPTERS 1–2)

1 The **cell** is the basic unit of life.

2 A **unicellular** organism consists of one cell which possesses a variety of structures enabling it to perform all the functions necessary for the maintenance of life.

3 A **multicellular** organism consists of more than one cell. In advanced animals and plants these are arranged into tissues giving a division of labour.

4 **Variation** in cell structure exists between cells of one type of tissue and cells of different tissues since cells are **specialised** to perform particular functions.

5 Cells absorb molecules in solution by **diffusion, osmosis** and **active transport**.

6 The **cell wall** surrounding plant cells is made of **cellulose** and is **freely permeable** to solutions. It prevents the cell from bursting when water is absorbed.

7 The **plasma membrane** surrounding the living contents of all cells is **selectively permeable**. It consists of **protein** and **phospholipid** molecules thought to be arranged as in the **fluid mosaic** model.

8 Tiny channels in the plasma membrane make it **porous** and allow the passive transport of small molecules by diffusion and osmosis along a **concentration gradient**.

9 Other chemical substances such as **ions** are actively transported across the plasma membrane against a concentration gradient by **protein carriers**. This process requires **energy**.

3 ATP and energy release

Effect of adenosine triphosphate (ATP) on muscle fibre

In the experiment shown in figure 3.1, only **ATP** is found to bring about contraction of the muscle fibres. It is therefore concluded that ATP is able to immediately provide the energy required for muscle contraction whereas glucose, despite being an energy-rich compound, is unable to do so.

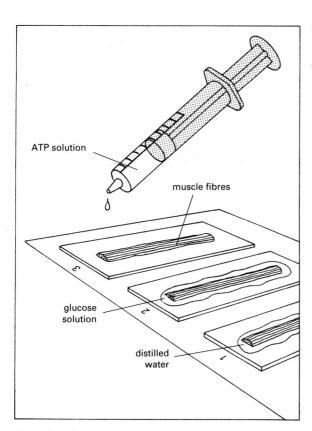

Figure 3.1   Investigating effect of ATP on muscle

Structure of ATP

A molecule of ATP is composed of adenosine and three inorganic phosphate (Pi) groups as shown in figure 3.2.

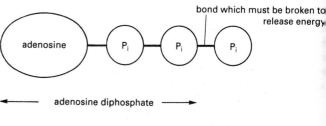

Figure 3.2   Structure of ATP

Energy stored in an ATP molecule is released when the bond attaching the terminal phosphate is broken by enzyme action. This results in the formation of adenosine diphosphate (ADP) and inorganic phosphate (Pi).

On the other hand, energy is required to regenerate ATP from ADP and inorganic phosphate by an enzyme-controlled process called **phosphorylation**.

This reversible reaction is summarised by the equation:

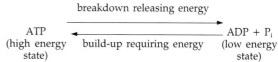

Production of ATP

When an energy-rich substance such as glucose is broken down (oxidised) in a living cell, it releases energy which is used to produce ATP.

When glucose is burned in a dish in the

laboratory, its energy is released in one quick burst of heat and light. However in a living cell, the breakdown of glucose during cell respiration (see chapter 4) is a *gradual* process involving many enzyme-controlled steps. This *orderly* release of energy is the ideal means by which the chemical energy needed to regenerate ATP from ADP + Pi is made available.

Importance of ATP

Many molecules of ATP are present in every living cell. Since ATP can rapidly revert to ADP + Pi, it is able to make energy available for energy-requiring processes (e.g. muscular contraction, protein synthesis, active transport of molecules and transmission of nerve impulses).

ATP is important because it acts as the *link* between energy-releasing reactions and energy-consuming reactions. It provides the means by which chemical energy is *transferred* from one type of reaction to the other in a living cell as shown in figure 3.3.

Since ATP breakdown is **coupled** with energy-demanding reactions, this promotes the transfer of energy to the new chemical bonds (e.g. those joining amino acids together) and helps to reduce the amount of energy that is lost as heat.

Turnover of ATP molecules

It has been estimated that an active cell (e.g. a bacterium undergoing cell division) requires approximately two million molecules of ATP per second to satisfy its energy requirements. This is made possible by the fact that a **rapid turnover** of ATP molecules constantly occurs in a cell. At any given moment some ATP molecules are undergoing breakdown and releasing the energy

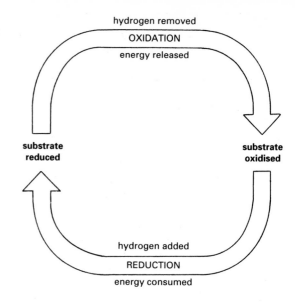

Figure 3.4   Oxidation and reduction

needed for cellular processes while others are being regenerated from ADP + Pi using energy released during cell respiration.

Oxidation and reduction

In a metabolic pathway, **oxidation** occurs when hydrogen ions are removed from the substrate and energy is released as shown in figure 3.4. Oxidation occurs during cell respiration (see chapter 4).

The process of **reduction** involves the addition of hydrogen to the substrate and the consumption of energy. Reduction occurs during photosynthesis (see chapter 6).

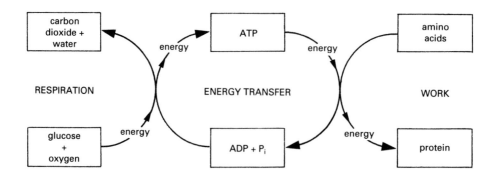

Figure 3.3   Transfer of chemical energy by ATP

QUESTIONS

1 a) What compound is represented by the letters ATP?

b) Give a word equation to indicate how ATP is regenerated in a cell.

c) Explain briefly the importance of ATP to a cell.

2 Metabolism (the sum of all the chemical changes that occur in a living organism) falls into two parts: **anabolism** which consists of energy-requiring reactions involving synthesis of complex molecules, and **catabolism** consisting of energy-yielding reactions in which complex molecules are broken down. **ATP transfers energy** from catabolic reactions to the anabolic ones.

Figure 3.5 is based on the information in the above paragraph.

a) Copy the diagram and add four arrow heads to show the directions in which the two coupled reactions occur.

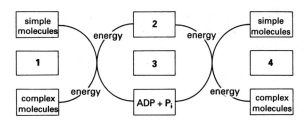

Figure 3.5

b) Complete boxes 1–4 using each of the terms given in bold print in the passage.

3 The human body produces ATP at a rate of approximately 400 g/hour, yet at any given moment there is only about 50 g present in the body. Explain why.

4 What would be the effect of depriving healthy respiring cells of inorganic phosphate? Explain your answer.

5 Give TWO differences between the processes of oxidation and reduction.

4 Chemistry of respiration

Respiration is the process by which chemical energy is released from a foodstuff by **oxidation**. It occurs in every living cell and involves the regeneration of the high energy compound ATP by a complex series of metabolic reactions.

Glycolysis

In the cytoplasm of a living cell, the process of cell respiration begins with a molecule of **6-carbon glucose** being broken down by a series of enzyme-controlled steps to form two molecules of **3-carbon pyruvic acid** (see figure 4.1).

This process of 'glucose-splitting' is called **glycolysis**. It requires energy from two molecules of ATP to trigger it off but later in the process sufficient energy is released to form four molecules of ATP giving a **net gain of 2 ATP**.

During glycolysis, hydrogen released from the respiratory substrate becomes temporarily bound to a **coenzyme** molecule. (At no point in the pathway does hydrogen exist as free atoms or molecules).

The process of glycolysis does not require oxygen but the hydrogen bound to a reduced coenzyme only produces further molecules of ATP (at a later stage in the process) if oxygen is present. In the absence of oxygen, **anaerobic respiration** occurs (see page 24).

Mitochondria

When oxygen is present, **aerobic respiration** occurs in the cell's **mitochondria** (see figure 4.2).

These are sausage-shaped organelles present in the cytoplasm of living cells. The inner membrane of each mitochondrion is folded into many plate-like extensions (**cristae**) which give a large surface area on which the processes of respiration can take place. The cristae protrude

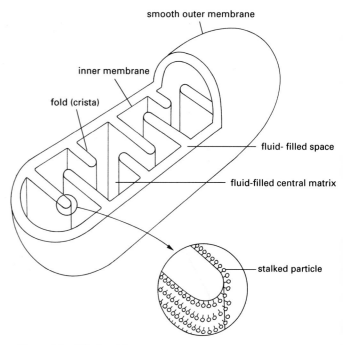

Figure 4.2   Mitochondrion

into the fluid-filled interior (**matrix**) which contains enzymes.

Electron micrographs reveal that each crista bears many **stalked particles**. These are the site of ATP production. Cells requiring a lot of energy, such as sperm, liver, muscle and nerve cells, contain numerous large mitochondria with many cristae.

Fate of pyruvic acid

Pyruvic acid produced during glycolysis diffuses into the matrix of a mitochondrion. Here it is converted into a **2-carbon compound**. This reaction is accompanied by the release of hydrogen which again becomes bound to a coenzyme.

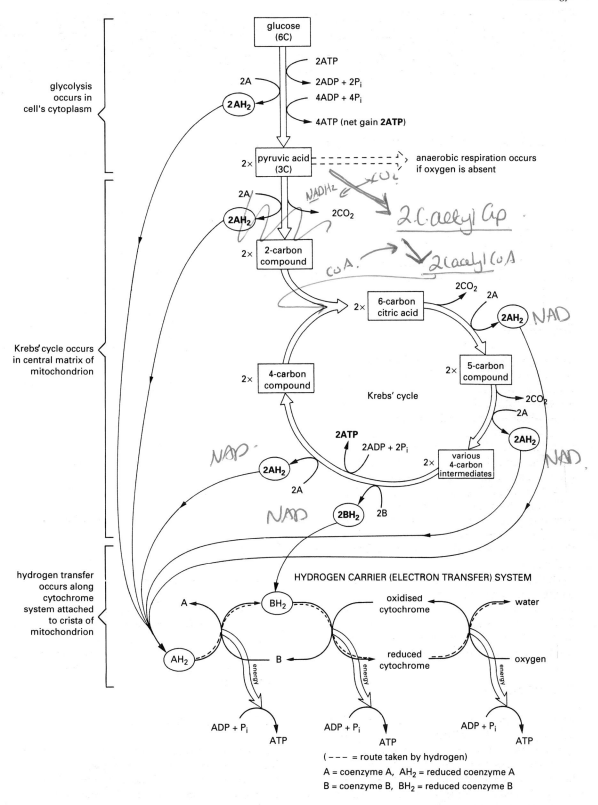

Figure 4.1   Chemistry of respiration

Krebs' cycle

This aerobic phase of respiration is also known as the **Citric Acid Cycle** and the **Tricarboxylic Acid Cycle**.

Each molecule of the 2-carbon compound (formed from pyruvic acid) reacts with a molecule of a **4-carbon compound** present in the matrix of the mitochondrion to form **6-carbon citric acid**. This is gradually converted back to the 4-carbon compound by a cyclic series of enzyme-controlled reactions (see figure 4.1).

One of these reactions brings about the **direct synthesis** of two molecules of **ATP**. The others are involved in the release of carbon and hydrogen from the respiratory substrate (e.g. glucose).

Enzymes controlling the release of carbon to form carbon dioxide are called **decarboxylases**. The carbon dioxide formed diffuses out of the cell as a waste product. Enzymes controlling the release of hydrogen are called **dehydrogenases**.

Hydrogen transfer

Figure 4.1 shows that there are six points along the pathway where hydrogen is released and becomes temporarily bound to one of the two types of coenzyme molecule. For the sake of simplicity these coenzyme molecules will be referred to as A and B. 'A' becomes reduced to AH_2 and 'B' becomes reduced to BH_2.

These reduced coenzymes transfer hydrogen to a chain of hydrogen carriers called the **cytochrome system**. Each mitochondrion possesses many of these systems, each attached to a crista.

Transfer of hydrogen from AH_2 along the cytochrome system releases sufficient energy to produce 3 ATP. Transfer of hydrogen from BH_2 (which enters the system further along) releases sufficient energy to produce 2 ATP.

In each case this process is called **oxidative phosphorylation** and in total generates 34 ATP per molecule of glucose. Thus the complete oxidation of one glucose molecule yields a total of **38 ATP** as explained in table 4.1. (Since the hydrogen carrier system produces 34 of the 38 ATP, it is the most important means of releasing energy during respiration.)

Figure 4.1 shows that the *final* hydrogen acceptor is **oxygen**. Hydrogen and oxygen combine under the action of the enzyme **cytochrome oxidase** to form **water**. Although oxygen only plays its part at the very end of the pathway, its presence is essential for hydrogen

source	location	number of ATP formed
direct synthesis during glycolysis	cytoplasm	2 (net gain)
2AH$_2$ formed during glycolysis	cytoplasm	6 (2× 3)
2AH$_2$ formed from breakdown of pyruvic acid	mitochondrion	6 (2 × 3)
6AH$_2$ formed during Krebs' cycle	mitochondrion	18 (6 × 3)
2BH$_2$ formed during Krebs' cycle	mitochondrion	4 (2 × 2)
direct synthesis during Krebs' cycle	mitochondrion	2
	TOTAL =	**38 ATP**

Table 4.1   Origin of ATP from aerobic breakdown of one molecule of glucose

to pass along the cytochrome system. In the absence of oxygen, the oxidation process cannot proceed beyond glycolysis.

Alternative respiratory substrate

Fatty acids are converted in the matrix of a mitochondrion to the same 2-carbon compound produced from pyruvic acid described earlier. This compound then enters the Krebs' cycle and produces ATP as before. Fat liberates more than double the energy released by the same mass of carbohydrate.

Amino acids from protein can also act as respiratory substrates. Alanine, for example, is converted to pyruvic acid so that it can enter the pathway and release energy. A certain amount of energy is always derived from excess dietary protein but tissue protein is only used as a source of energy during prolonged starvation.

Function of aerobic respiration

Aerobic respiration (summarised in figure 4.3) is a metabolic pathway with a series of enzyme-controlled reactions by which a respiratory substrate such as 6-carbon glucose is oxidised to form carbon dioxide accompanied by the production of ATP from ADP + Pi.

This **regeneration** of ATP for use in other cellular processes is the key function of respiration. (ATP is also regenerated in photosynthesis by **photophosphorylation**, see page 33.)

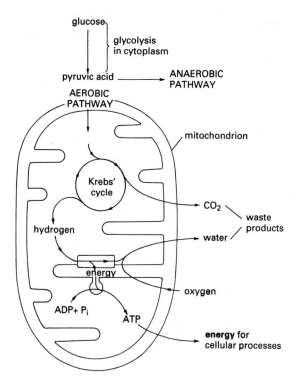

Figure 4.3 Summary of chemistry of respiration

Dehydrogenase activity in yeast

During respiration, glucose is gradually broken down (oxidised) and hydrogen is released at various stages along the pathway. Each of these stages is controlled by an enzyme called a **dehydrogenase**.

Resazurin dye is a chemical which changes colour when it is reduced (i.e. gains hydrogen) as follows:

blue	→	pink	→	colourless
(unreduced)		(partially reduced)		(reduced)

Before setting up the experiment shown in figure 4.4, dried yeast is added to water and aerated for an hour at 35°C to ensure that the yeast is in an active state.

The contents of tube **A** are found to change from blue via pink to colourless much faster than those in tube **B**. Tube **C**, the control, remains unchanged.

It is concluded that, in tube **A**, hydrogen has been rapidly released and has reduced the resazurin dye. For this to be possible, **dehydrogenase enzymes** present in the yeast cells must have acted on glucose, the respiratory substrate, and **oxidised** it.

In tube **B**, the reaction was slower since no glucose was added and the dehydrogenases could only act on any small amount of respiratory substrate already present in the yeast cells.

In tube **C**, boiling has killed the cells and denatured the dehydrogenase enzymes.

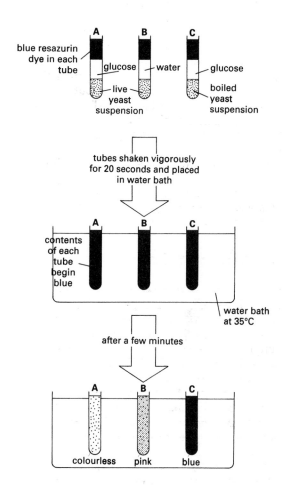

Figure 4.4 Dehydrogenase activity

Effect of a heavy metal (lead) on catechol oxidase

When a potato or an apple is peeled, its inner white tissues turn brown if left exposed to oxygen. This browning effect is brought about by the enzyme **catechol oxidase** (polyphenol oxidase) which is present in the cells and catalyses the reaction below:

$$\text{phenols + oxygen} \xrightarrow[\text{oxidase}]{\text{catechol}} \text{yellow-brown + water}$$

(e.g. catechol)   compounds

23

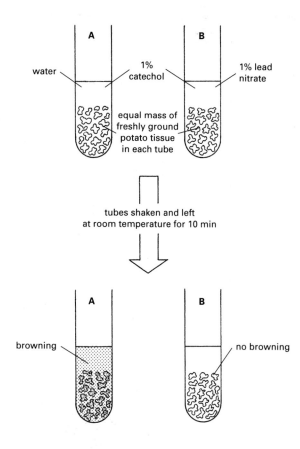

Figure 4.5   Effect of lead nitrate on catechol oxidase

Look at the experiment shown in figure 4.5. In test tube **A**, finely ground potato tissue quickly turns brown as the enzyme from the cells acts on catechol (the substrate). However, in tube **B**, the potato tissue fails to turn brown in the presence of lead nitrate.

It is therefore concluded that **lead** (a heavy metal) inhibits the activity of the enzyme, catechol oxidase. Similarly heavy metals can inhibit many of the enzymes which control respiratory pathways (see also pages 162–163).

Measuring rate of respiration

Figure 4.6 shows a simple **respirometer** set up to measure an earthworm's rate of respiration (as volume of oxygen consumed per unit time). Carbon dioxide given out by the animal is absorbed by the sodium hydroxide. Oxygen taken in by the animal causes a decrease in volume of the enclosed gas, so the coloured liquid rises up the tube.

After a known length of time (e.g. one hour) the syringe is used to find out the volume of air which must be injected to return the coloured liquid to its initial level. If for example 0.4 ml is needed, then the worm's respiratory rate under these conditions = 0.4 ml oxygen/hour.

The control is set up without an earthworm. It could be argued that the observed differences were simply due to the fact that some space was occupied in the respirometer and not necessarily caused by the worm respiring. The best control includes some inert material (such as glass beads) equal in volume to the animal.

The experiment is carried out with the respirometer and control tubes in a large container of water at room temperature to ensure constant temperature throughout.

Anaerobic respiration

This is the process by which a little energy is derived from the **partial breakdown** of sugar in the absence of oxygen. Since oxygen is unavailable to the cell, the hydrogen transfer system and the Krebs' cycle (see figure 4.1) cannot operate in any of the cell's mitochondria. Only glycolysis can occur.

Each glucose molecule is partially broken down to pyruvic acid and yields only two molecules of ATP. The hydrogen released cannot go on to make ATP in the absence of oxygen. The pyruvic acid undergoes one of the following metabolic pathways (depending on the organism involved).

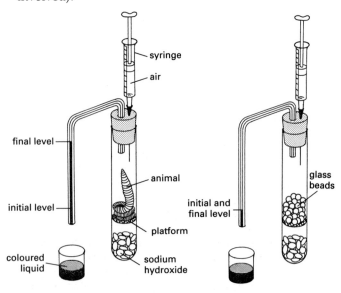

Figure 4.6   Simple respirometer and control

Anaerobic respiration in plants

The equation below summarises this process in plant cells such as yeast and water-logged roots:

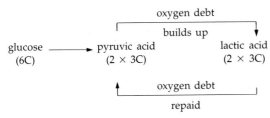

glucose   ⟶   pyruvic acid ⟶ ethanol + $2CO_2$
(6C)   (2 × 3C)   (2 × 2C)

Anaerobic respiration in animals

The equation below summarises this process in animal cells such as skeletal muscle tissue:

oxygen debt
builds up
glucose   ⟶   pyruvic acid   lactic acid
(6C)   (2 × 3C)   (2 × 3C)
oxygen debt
repaid

During lactic acid formation, the body accumulates an **oxygen debt**. This is repaid when oxygen becomes available and lactic acid is converted back to pyruvic acid which then enters the aerobic pathway.

Anaerobic respiration is a less efficient process since it produces only **2 ATP** per molecule of glucose compared with **38 ATP** formed by aerobic respiration. The majority of living cells thrive in oxygen and respire aerobically. They only resort to anaerobic respiration to obtain a little energy for survival while oxygen is absent.

process	site where process occurs in cell	reaction(s) involved in process	products
A glycolysis	[_____]	splitting of [_____] into 3-carbon compound	ATP and [____ ____]
B [_____] cycle	central matrix of [_____]	removal of [_____] from carbon compounds by oxidation; release of carbon atoms forming [____ ____]	carbon dioxide
C [_____] (hydrogen acceptor) system	[_____] of mitochondrion	release of energy from hydrogen (oxidation of hydrogen)	[_____] and [_____]

Table 4.2

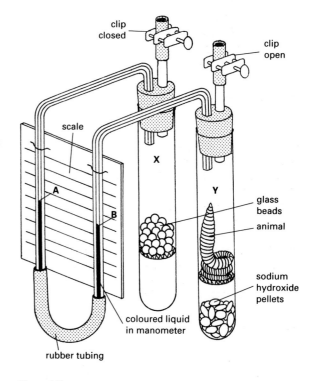

Figure 4.7

QUESTIONS

1 State ONE way in which the structure of a mitochondrion is ideally suited to its function.

2 Table 4.2 refers to the process of cell respiration.
   a) Copy the table and complete the blanks indicated by brackets.
   b) During which process in the table do the products of fat digestion enter the pathway and begin to undergo oxidation?
   c) State which process(es) would fail to occur in the absence of oxygen.

3 Compare aerobic and anaerobic respiration in yeast cells with respect to ATP production and final products resulting from each process.

4 Give TWO reasons why a drop in pH occurs in the skeletal muscle tissue of a human during intensive physical training.

5 Figure 4.7 shows a set of apparatus about to be used to investigate the effect of temperature on an earthworm's rate of respiration.
   a) State TWO ways in which the apparatus would have to be altered before beginning the experiment.

b) Suggest how a temperature of 25°C could be obtained simultaneously in tube **X** and **Y**.

c) Assume that the experiment is correctly set up and running. Predict the direction in which levels **A** and **B** will now move. Explain why.

d) In what way will the movement of the liquid levels differ when the experiment is repeated at 5°C for the same length of time? Explain why.

e) Why should the same earthworm be used each time?

f) What is the purpose of the glass beads in tube **X**?

g) How could the apparatus be redesigned to measure the actual volume of oxygen consumed per hour by the earthworm?

What you should know (CHAPTERS 3–4)

1 **ATP** is a high energy compound which is able to release and transfer **energy** when it is required for cellular processes.

2 ATP is regenerated from ADP and inorganic phosphate by the process of **phosphorylation** using energy released during respiration.

3 **Oxidation** involves the removal of hydrogen from a substrate and the release of energy; **reduction** involves the addition of hydrogen to a substrate and the consumption of energy.

4 **Glycolysis** is a biochemical pathway common to aerobic and anaerobic respiration. It involves the **breakdown of glucose** to **pyruvic acid** in the cytoplasm of a cell with the net gain of 2 ATP.

5 In the presence of oxygen, **aerobic** respiration occurs in the **central matrix** of mitochondria where the respiratory substrate is oxidised during the **Krebs' cycle** and **hydrogen** is released.

6 This hydrogen becomes temporarily bound to **coenzymes** which transfer it to the **cytochrome** system on the **cristae** of mitochondria where energy is released and used to form ATP.

7 As a result of **aerobic** respiration, one molecule of glucose yields **38 ATP. Water** and CO_2 are the final metabolic products.

8 In the absence of oxygen, **anaerobic** respiration occurs and one molecule of glucose yields **2 ATP**. The final metabolic products are **ethanol** and CO_2 in plant cells, and **lactic acid** in animal cells (and some bacteria).

5 Absorption of light by photosynthetic pigments

Absorption, reflection and transmission

Although some of the light striking a leaf (see figure 5.1) is **reflected** and some is **transmitted**, most of it is **absorbed**. Of this absorbed light, only a small part is used in **photosynthesis**. The rest is converted to heat and lost (e.g. by radiation).

Extraction of leaf pigments

Fresh leaves are finely chopped and ground in a mortar containing acetone and a little silver sand as shown in figure 5.2. The extract of soluble pigments is separated from the cell debris by filtration.

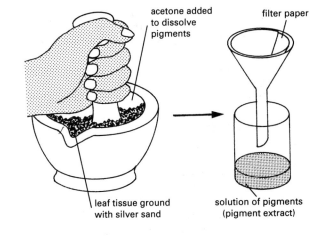

Figure 5.2   Extraction of leaf pigments

Separation of leaf pigments

A strip of chromatography paper is prepared as shown in figure 5.3. This diagram also illustrates the procedure followed during the spotting of the extract and use of the chromatography solvent.

Spotting and drying of the extract is repeated many times. The end of the paper is then dipped into the solvent. The **chromatogram** is allowed to run until the solvent has almost reached the top of the paper. Table 5.1 gives the reasons for adopting certain techniques and precautions during this experiment.

Chromatogram

Figure 5.4 shows the chromatogram formed as a result of this process of **ascending chromatography**. The four pigments become separated because each differs from the others in its degree of solubility in the solvent.

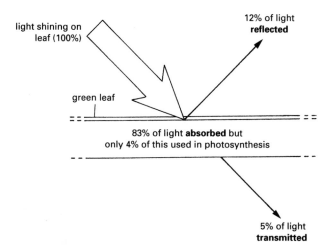

Figure 5.1   Fate of light striking a leaf

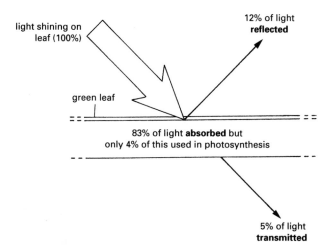

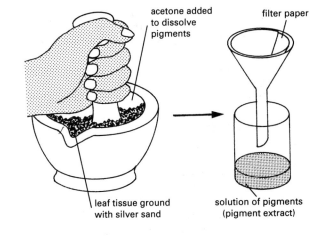

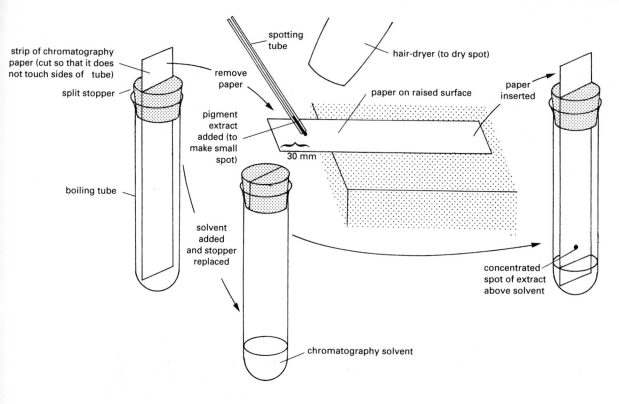

Figure 5.3   *Separation of leaf pigments*

design feature or precaution	reason
plant tissue ground in silver sand	to rupture cells allowing release of contents
chromatography paper cut so that it does not touch sides of tube	to ensure that solvent rises uniformly through paper rather than more rapidly up its edges
chromatography solvent kept in stoppered tube prior to use	to allow atmosphere inside tube to become saturated with vapour
chromatography paper placed on raised surface and spotting done on overlap	to stop spot spreading and to prevent it from becoming contaminated with dirt or any other chemical
spotting and drying repeated many times	to obtain a concentrated spot of pigments
paper positioned in tube so that pigment spot is above solvent level at start	to prevent extract dissolving in main bulk of solvent at bottom of tube
naked flames switched off during experiment	to prevent fire risk since acetone and petroleum ether are highly flammable

Table 5.1   *Design techniques*

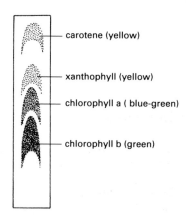

Figure 5.4   *Chromatogram of leaf pigments*

The solvent carries the most soluble (carotene) to the highest position and so on down the paper to chlorophyll b, the least soluble. This is carried the shortest distance. (A grey-coloured breakdown product of chlorophyll sometimes appears on the chromatogram but it plays no part in photosynthesis.)

Chlorophyll is a structurally complex molecule containing **magnesium**.

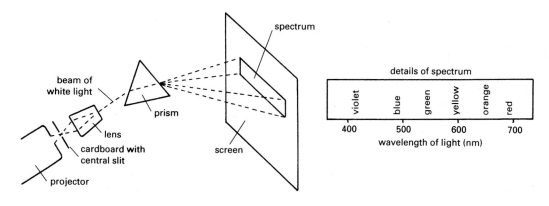

Figure 5.5   Spectrum of white (visible) light

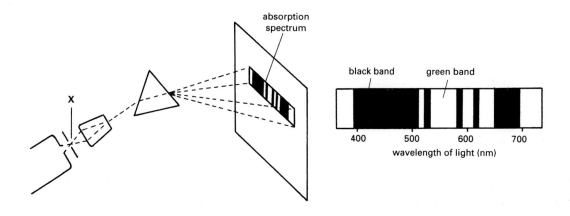

Figure 5.6   Absorption spectrum of leaf pigments

Absorption spectra

When a beam of light is passed through a glass prism, the **spectrum** of **white (visible) light** is produced (see figure 5.5).

When a beam of white light is first passed through a sample of leaf pigments placed at **X** in figure 5.6 and then passed through a glass prism, an **absorption spectrum** is produced.

Each black band is a region of the spectrum where light energy has been absorbed by the leaf pigments and has therefore failed to pass through the prism and onto the screen. Each coloured band (e.g. green) is a region where light has not been absorbed by the extract.

Since the wavelengths of light not absorbed by a pigment are transmitted or reflected, chlorophyll appears green to the eye.

Graphs of absorption and action spectra

At each wavelength the **degree of absorption** of visible light by each pigment can be measured using a **spectrometer**. The data obtained allows a detailed graph of each pigment's absorption spectrum to be plotted (see figure 5.7).

An **action spectrum** (see figure 5.8) charts the effectiveness of different wavelengths of light at bringing about the process of photosynthesis.

Comparison of figures 5.7 and 5.8 shows that a close correlation exists between the overall absorption spectrum for the pigments and the action spectrum.

It is therefore concluded that the absorption of certain wavelengths of light for use in photosynthesis is the crucial role played by these pigments.

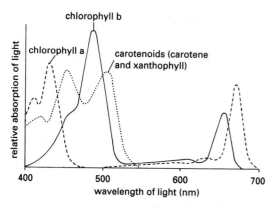

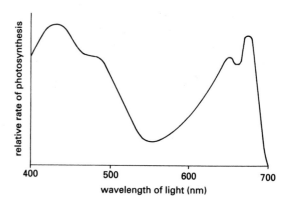

Figure 5.7 Graph of absorption spectra of leaf pigments

Figure 5.8 Action spectrum

The two chlorophylls absorb light primarily in the **red** and **blue** regions of the spectrum, but the accessory pigments (carotenoids) absorb light from other regions and pass the energy on to chlorophyll.

Seaweed pigments

In place of chlorophyll b, brown seaweeds possess chlorophyll c and red seaweeds possess chlorophyll d. In addition, brown seaweeds are rich in **fucoxanthin** (a form of the yellow pigment xanthophyll which masks the green chlorophyll). This pigment has an especially broad absorption spectrum and absorbs light in parts of the spectrum where chlorophyll is less efficient. Red algae contain a red pigment called **phycoerythrin** which absorbs green light.

These pigments extend the range of light wavelengths which can be absorbed and used for photosynthesis by brown and red seaweeds. This partly compensates for the lower intensity of light which reaches these plants in their dimly lit rock pool and deep water habitats.

Green seaweeds, which are typical of shallower water and brightly lit rock pools, lack fucoxanthin and phycoerythrin.

Chloroplast

Chloroplasts (figure 5.9) are relatively large, **discus-shaped** organelles situated in the cytoplasm of green plant cells. Each is bounded by a double membrane and possesses distinct internal structures made of **grana** and **lamellae**.

Each granum consists of a coin-like stack of flattened sacs containing **photosynthetic pigments**. Grana are the site of the light-dependent stage of photosynthesis (see page 33). Lamellae are tubular extensions which form an interconnecting network between grana but do not contain chlorophyll.

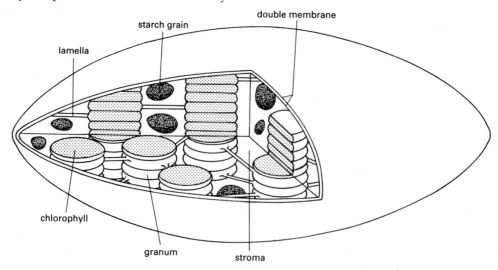

Figure 5.9 Chloroplast

The colourless background material in a chlorophast is called **stroma**. It is the site of the carbon fixation stage of photosynthesis (see page 33). It lacks chlorophyll but contains important enzymes and starch grains (which act as temporary stores of photosynthetic products).

QUESTIONS

1 Some of the light that strikes a leaf is transmitted.
   a) What does this mean?
   b) State TWO possible fates of the light that is not transmitted.
2 Figure 5.10 shows the result of placing a strand of alga in a liquid containing motile aerobic bacteria and illuminating the strand with a tiny spectrum of light.

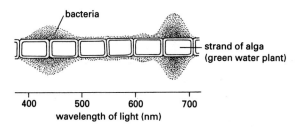

Figure 5.10

   a) In which colours of light did most bacteria congregate?
   b) Account for this distribution of the bacteria.
3 Why does chlorophyll b appear green in colour and phycoerythrin appear red?
4 Explain the difference between an absorption spectrum and an action spectrum.
5 a) Give TWO structural differences between the grana and the stroma found in a chloroplast.
   b) Identify the stage of photosynthesis that occurs in each of these regions.

EXPERIMENTAL DESIGN QUESTION

Observation: In addition to being able to use blue light (and to a lesser extent red light) for photosynthesis, a certain red seaweed is also found to photosynthesise efficiently in yellow-green light.

Hypothesis: This red seaweed contains one or more light-absorbing pigments not present in the leaves of a normal land plant such as dandelion.

Problem: Outline an experimental procedure that could be employed to test the validity of the hypothesis.

DATA INTERPRETATION

The information in table 5.2 refers to photosynthetic pigments. The data in table 5.3

class of pigment	name	principal (P) or accessory (A) pigment	location
chlorophyll	a	P	all photosynthetic plants
	b	A	higher plants and green algae
	c	A	brown algae
	d	A	some red algae
carotenoid	xanthophyll	A	all photosynthetic plants
	carotene	A	all photosynthetic plants
phycobilin	phycocyanin	A	main phycobilin in blue-green algae
	phycoerythrin	A	main phycobilin in red algae

Table 5.2

wavelength of light (nm)	amount of light absorbed (arbitrary units)	
	pigment 1	pigment 2
400	0	0
420	0	0
440	0	0
460	0.1	0
480	0.3	0.05
500	1.0	0.1
520	3.2	0.2
540	4.9	0.3
560	5.0	0.5
580	2.7	0.9
600	0.1	1.7
620	0.05	2.6
640	0	2.8
660	0	2.2
680	0	0.3
700	0	0

Table 5.3

refers to the amount of light absorbed at various wavelengths of light by two of these pigments each extracted from a different seaweed. All seaweeds belong to a large group of plants called algae.

a) Name ONE principal and THREE accessory pigments found in brown seaweeds. (1)

b) Name a type of plant that would lack all four classes of pigment listed in table 5.2. (1)

c) Plot the data given in table 5.3 as two line graphs (curves) to show the absorption spectra of pigments 1 and 2. (3)

d) Refer back to figure 5.5 on page 29 which shows the spectrum of white (visible) light and devise a way of adding the six colours to one of the axes in your graph. (1)

e) Pigments 1 and 2 are known as the phycobilins.

(i) One of them absorbs orange and red light and appears blue-green to the eye. Identify it by its number and its proper name.

(ii) One of them absorbs green and yellow light and appears red to the eye. Identify it by its number and its proper name. (2)

f) Imagine that a sample of each of the phycobilin pigments were placed in turn at point X in the experiment shown in figure 5.6 on page 29. Make a simple diagram of the absorption spectrum that you would expect in each case.

(1)

g) Phycoerythrin is commonly found in seaweeds that live in deep sea water or in dimly lit rock pools. Suggest how the presence of this pigment helps the plant to survive. (1)

(10)

6 Chemistry of photosynthesis

Photosynthesis is the process by which organic compounds are synthesised by the **reduction** of carbon dioxide. The energy required for this process comes from light energy. The light energy is absorbed by photosynthetic pigments.

Photosynthesis consists of two separate parts: a light-dependent (photochemical) stage called **photolysis** and a temperature-dependent (thermochemical) stage called **carbon fixation**.

Photolysis

This light-dependent stage occurs in the grana of chloroplasts (see figure 5.9 on page 30). Solar (light) energy is trapped by chlorophyll (and accessory pigments) and is converted into **chemical energy**. This process involves several important events which are summarised in figure 6.1.

Light energy is used to split molecules of water into hydrogen and oxygen. This is called the **photolysis of water**. The oxygen is released as a by-product. The hydrogen combines with a **hydrogen acceptor** (which we will refer to as D) to form reduced hydrogen acceptor (DH_2).

In addition, chlorophyll makes energy available for the regeneration of ATP from ADP and inorganic phosphate. This process is called **photophosphorylation**.

The hydrogen held by DH_2 and the energy held by ATP at the *end* of the light stage are essential for use in **carbon fixation**, the second stage of photosynthesis.

Carbon fixation

This thermochemical stage occurs in the stroma of a chloroplast. It consists of several enzyme-controlled chemical reactions which take the form of a cycle (often referred to as the **Calvin cycle** after the scientist who discovered it). Figure 6.2 summarises the cycle and indicates the number of carbon atoms in the molecules of the metabolites involved.

On entering a chloroplast by diffusion, a molecule of carbon dioxide combines with a molecule of **5-carbon ribulose bisphosphate (RuBP)** (also known as ribulose diphosphate (RuDP)) to form a 6-carbon molecule. This molecule is unstable and rapidly splits into two molecules of **3-carbon glycerate–3-phosphate (GP)** (also known as phosphoglyceric acid (PGA)).

In the next stage of the cycle, GP is converted to a **3-carbon sugar (triose phosphate)**. This conversion requires using the hydrogen temporarily bound to the reduced hydrogen acceptor (DH_2) and some of the energy held in ATP. (Both of these are provided by the light-dependent stage.)

Each molecule of **glucose** (a **6-carbon/hexose sugar**) is synthesised from a pair of these 3-

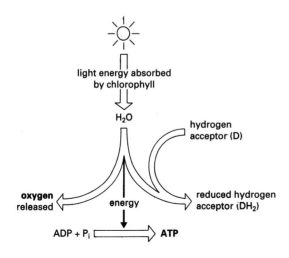

Figure 6.1   Light-dependent stage (photolysis)

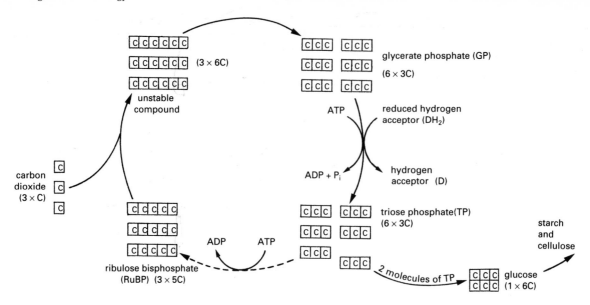

Figure 6.2   Calvin cycle (carbon fixation)

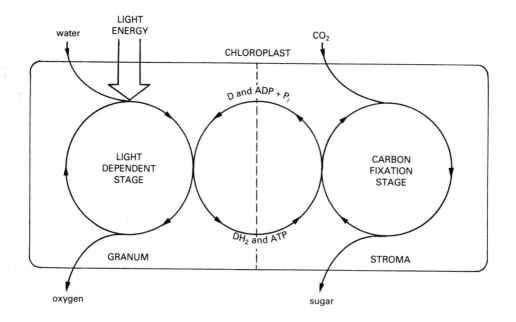

Figure 6.3   Summary of photosynthesis

carbon sugar molecules which combine together in an enzyme-controlled sequence of reactions. Molecules of glucose may then be built up into **starch** and **cellulose**.

However, the 3-carbon triose phosphate molecules are not all used to make complex products. Some are needed to regenerate RuBP, the carbon dioxide acceptor. This conversion of triose phosphate (5 × 3C) to RuBP (3 × 5C) also requires energy which is derived from ATP.

Summary

The process of photosynthesis involves the **fixation of energy** and is summarised in figure 6.3.

Reduction is the process by which hydrogen is added to a substrate (see page 18). Photosynthesis is therefore a reduction process since carbohydrate is formed by the reduction of carbon dioxide.

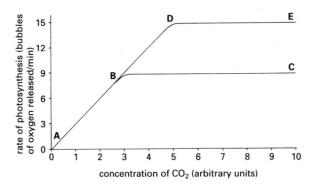

Figure 6.4   Limiting factors

Limiting factors

Photosynthesis is affected by several factors. These include **temperature, light intensity** and **carbon dioxide concentration**.

The rate at which photosynthesis proceeds is limited by whichever of these factors is in short supply. For example light intensity would probably be the factor limiting photosynthesis on a dull wet summer's day.

The principle of **limiting factors** is illustrated by figure 6.4. Graph **ABC** refers to a water plant kept in conditions of constant low light intensity. The carbon dioxide concentration is varied by adding a chemical to the water. Photosynthetic rate is measured by counting the number of oxygen bubbles released per minute.

When the plant is supplied with a CO_2 concentration of only 1 unit, photosynthetic rate is limited (by this low concentration of CO_2) to 3 oxygen bubbles/minute. When the CO_2 concentration is increased to 2 units, photosynthetic rate increases to 6 oxygen bubbles/minute but no further since CO_2 concentration becomes limiting again. A further increase in CO_2 concentration to 3 units brings about a further increase in photosynthetic rate. Beyond this point, the graph levels out and any further increases in CO_2 concentration do not affect photosynthetic rate. Light (which has been at constant low intensity throughout) has now become the factor limiting the process.

Graph **ADE** represents a further experiment using the same plant kept in conditions of constant high light intensity. This time an increase in CO_2 concentration to 4 and 5 units bring about a corresponding increase in photosynthetic rate in each case because, at these concentrations, CO_2 is still the limiting factor when light intensity is high. Beyond 5 units of

CO_2, the graph levels off again since light intensity (or some other factor) has become limiting.

QUESTIONS

1 **a)** What happens to a molecule of water during the light reaction?
 b) What is meant by the term photophosphorylation?
2 **a)** Name the substances represented by the letters GP and RuBP in figure 6.5.

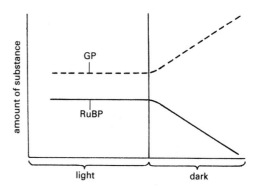

Figure 6.5

 b) Describe the relationship that is thought to exist between GP and RuBP when light is present.
 c) Briefly explain how the results shown in the graph supply evidence that this relationship does exist.
3 Figure 6.6 represents the carbon fixation stage of photosynthesis.
 a) Copy and complete the diagram by inserting the names of the appropriate intermediate metabolites in the boxes.
 b) Which compounds in the diagram are the products of the light-dependent reaction?
 c) Which raw material shown in the diagram is essential for the Calvin cycle to turn, but plays no part in the light-dependent reaction?
4 Figure 6.7 shows the effect of increasing light intensity on the rates of photosynthesis of a plant kept at two different concentrations of carbon dioxide.
 a) State TWO ways in which rate of photosynthesis can be measured.
 b) What factor was limiting photosynthetic rate at region **A** on the graph?
 c) What factor was limiting photosynthetic rate at point **B**?

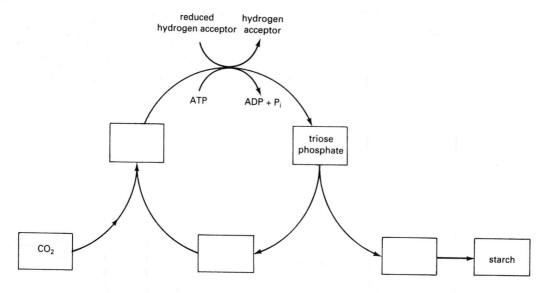

Figure 6.6

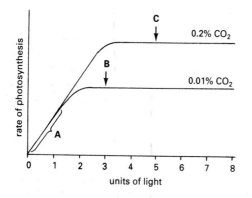

Figure 6.7

d) By how many times did CO_2 concentration differ between the two experiments?

e) What experimental factor, not referred to in the graph, could be limiting the rate of photosynthesis at point **C**?

What you should know (CHAPTERS 5–6)

1 Light is **absorbed, reflected** and **transmitted** by a leaf.
2 **Chlorophyll** absorbs light primarily in the **blue** and **red** regions of the spectrum of white light.
3 **Yellow** pigments absorb **blue-green** light.
4 **Chloroplasts** possess internal structures called **grana** which contain photosynthetic pigments and are the site of the **light-dependent stage** of photosynthesis.
5 The region between grana is called **stroma**. It is the site of the **carbon fixation** stage of photosynthesis.

6 The light-dependent stage of photosynthesis is called **photolysis**. It produces the energy (held in **ATP**) and **hydrogen** needed for the second stage (carbon fixation).
7 The second stage consists of a **cycle** of reactions which brings about the **reduction of carbon dioxide** using the ATP and hydrogen from photolysis to form **carbohydrate**.
8 Photosynthesis is affected by temperature, light intensity and carbon dioxide concentration. Its rate is therefore **limited** by whichever one of these factors is in short supply.

7 DNA and its replication

Genetic material of a cell

Chromosomes are thread-like structures found in the nucleus of a cell. They are composed of protein and **deoxyribonucleic acid** (**DNA**). **Genes**, which determine an organism's characteristics, are lengths of DNA.

Structure of DNA

A molecule of DNA consists of two strands, each made up of repeating units called **nucleotides** (see figure 7.1). There are four different **bases** (**adenine, thymine, guanine** and **cytosine**), and so four different types of nucleotide exist.

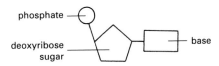

Figure 7.1   Structure of a nucleotide

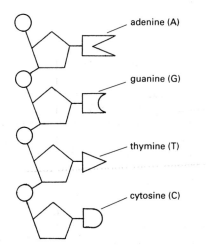

Figure 7.2   Strand of nucleotides

Bonds between phosphate and **deoxyribose** sugar molecules join neighbouring nucleotide units together into a strand (see figure 7.2). Two of these strands become joined together by weak (hydrogen) bonds between their bases. However each base can only join with one other type: adenine (A) always bonds with thymine (T), and guanine (G) always bonds with cytosine (C). A-T and G-C are called **base pairs**.

The resultant double-stranded molecule is DNA and its two strands are arranged as shown in figure 7.3. This twisted coil is called a **double helix**. It is like a spiral ladder in which the sugar-phosphate 'backbones' form the uprights and the base pairs form the rungs.

Replication of DNA

DNA is a unique molecule because it is able to reproduce itself *exactly*. This process is called **replication** and it is the means by which new genes are made. For replication to occur, the nucleus must contain the following:

- DNA (to act as a template for the new molecule);

- a supply of the four types of nucleotide;

- the appropriate enzymes (e.g. DNA polymerase);

- ATP (for energy).

The process of replication is illustrated in figure 7.4. It continues until the entire DNA molecule has become replicated. As each new molecule of DNA forms, it coils itself into a double helix. So DNA replication results in the formation of *two* new molecules, each of which receives *one* strand of the original parent molecule. It is therefore said to be **semi-conservative** (see figure 7.5).

The whole process occurs in the nucleus of a cell during interphase, immediately prior to

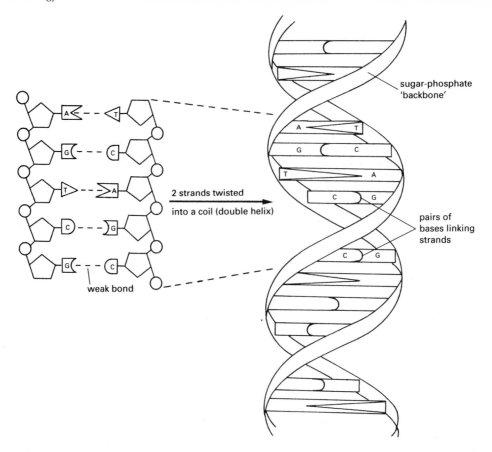

Figure 7.3 Structure of DNA

mitosis and meiosis (see page 60). At this time, the DNA content of a cell is at its maximum.

DNA replication ensures that an exact copy of the genetic information is passed from cell to cell during growth, and from generation to generation during reproduction. It is essential for the continuation of life.

QUESTIONS

1 Figure 7.6 shows part of a DNA strand.
a) Of which molecules is region **X** composed?
b) Draw the strand that would be complementary to the one shown in the diagram.

Figure 7.6

2 Draw a diagram to show two complementary nucleotides in opposite strands of a DNA molecule.

3 Calculate the percentage of thymine molecules present in a DNA molecule containing 1000 bases of which 200 are guanine.

4 List the substances that must be present before DNA replication can take place.

5 Place the following events that occur during DNA replication in the correct order:
a) bonds between opposite bases break;
b) bonds between opposite bases form;
c) DNA molecule uncoils from one end;
d) opposite strands of DNA separate;
e) daughter DNA molecules coil into double helices;
f) sugar-phosphate bonds form.

6 What proportion of a parent DNA molecule is contained in each daughter molecule after replication?

7 Table 7.1 gives the relative amounts of DNA present in some cell types from four different animals.

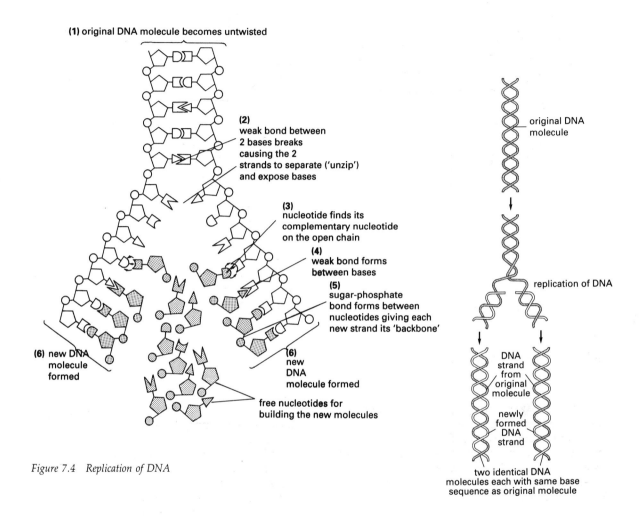

(1) original DNA molecule becomes untwisted

(2) weak bond between 2 bases breaks causing the 2 strands to separate ('unzip') and expose bases

(3) nucleotide finds its complementary nucleotide on the open chain

(4) weak bond forms between bases

(5) sugar-phosphate bond forms between nucleotides giving each new strand its 'backbone'

(6) new DNA molecule formed

(6) new DNA molecule formed

free nucleotides for building the new molecules

Figure 7.4   Replication of DNA

original DNA molecule

replication of DNA

DNA strand from original molecule

newly formed DNA strand

two identical DNA molecules each with same base sequence as original molecule

Figure 7.5   Semi-conservative replication

	sperm	red blood cell	kidney
carp	1.6	3.5	3.3
chicken	1.3	2.3	2.4
cow	3.3	0	6.4
human	3.3	0	6.6

Table 7.1

a) Account for the fact that the DNA content of human red blood cells is zero.

b) Based on the data, suggest a structural difference that exists between the red blood cells of a cow and a chicken.
c) Make a generalisation about the DNA content of sperm compared with kidney cells.
d) When kidney cells are removed from a young animal and grown in a tissue culture, their DNA content is found at times to be higher than the values given in the table. Explain why.

8 RNA and protein synthesis

The genetic code

The 'blueprint' for an organism's inherited characteristics is contained in the genetic instructions stored in its DNA. This **genetic code** consists of a sequence of bases which is unique to the organism.

The organism's characteristics are the result of many biochemical processes controlled by enzymes. Each enzyme is made of **protein** and its exact nature depends on the sequence of **amino acids** present in its polypeptide chains (see page 44). DNA therefore determines the organism's inherited characteristics by determining which enzymes are produced.

DNA possesses only four different bases yet proteins contain about twenty different amino acids. Clearly the relationship cannot be one base coding for one amino acid since this would only allow four amino acids to be coded. Even two bases per amino acid would give only sixteen (4^2) different codes. However if the bases are taken in groups of *three* then this gives 64 (4^3) different combinations (see also Appendix 4).

It is now known that each amino acid is coded for by one (or more) of these 64 **triplets** of bases. This is called the **triplet code**.

RNA

The second type of nucleic acid is called **ribonucleic acid** (RNA). It is a large molecule similar in structure to a single strand of DNA but containing the base **uracil** in place of thymine, and **ribose** sugar in place of deoxyribose (see figure 8.1). Two types of RNA are **messenger RNA (mRNA)** and **transfer RNA (tRNA)**.

Protein synthesis occurs in tiny structures called **ribosomes** (see page 42) situated in the cytoplasm. The genetic information needed for the construction of a protein with the correct

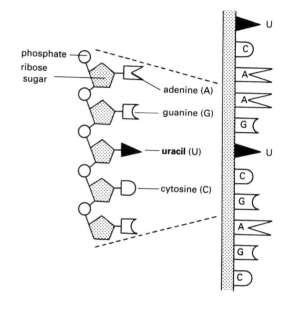

Figure 8.1   Structure of RNA

sequence of amino acids is carried from the DNA of a gene in the nucleus to a ribosome by a 'template' molecule of mRNA.

Protein synthesis

Transcription of DNA into mRNA.

The appropriate region of a DNA molecule (at the gene to be coded) temporarily splits open to expose its bases, as shown in figure 8.2. A molecule of mRNA is formed (**transcribed**) from one of the DNA strands using free nucleotides present in the nucleus. **Complementary base pairing** (G-C, C-G, T-A and A-U) determines the sequence of the bases on the mRNA molecule. Each triplet of bases on mRNA is called a **codon**.

Figure 8.2 Transcription of DNA into mRNA

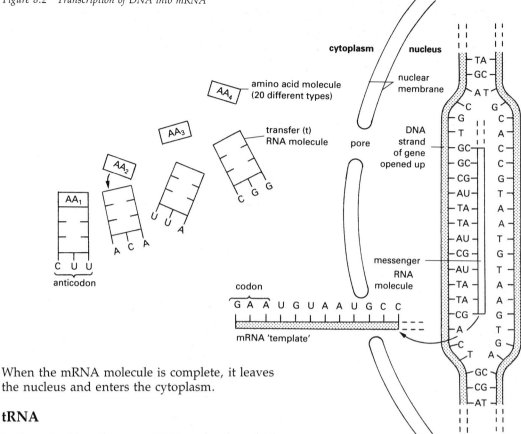

When the mRNA molecule is complete, it leaves the nucleus and enters the cytoplasm.

tRNA

In the cytoplasm there are tRNA molecules which each bear a triplet (**anticodon**) of bases corresponding to a *particular* amino acid. Each tRNA molecule picks up the appropriate amino acid from the cytoplasm. Every cell has as many types of tRNA as there are types of amino acid.

Translation of RNA into protein

The mRNA molecule becomes attached to a ribosome. Its triplet code is read and a tRNA anticodon links with each complementary codon of bases on the mRNA molecule (see figure 8.3).

Peptide bonds form between adjacent amino acids. The completed **protein** (consisting of very many amino acids) is then released into the cytoplasm.

Each tRNA molecule becomes attached to another molecule of its amino acid ready to repeat the process. The mRNA is often reused to produce further molecules of the same protein.

Thus each gene codes for one protein (or **polypeptide**) with the order of the amino acids in the protein being determined by the sequence of the bases in the gene's DNA. It is by this means that genes control an organism's phenotype. Variation amongst the members of a

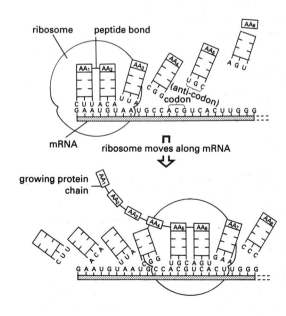

Figure 8.3 Translation of RNA into protein

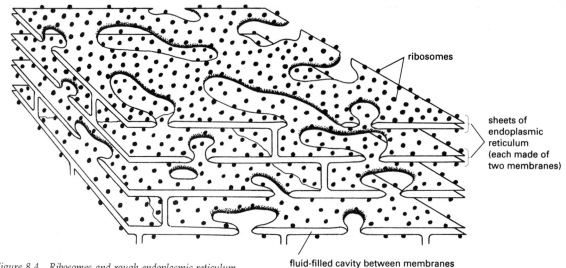

Figure 8.4 *Ribosomes and rough endoplasmic reticulum*

species results from differences in the sequence of the DNA's bases. These are brought about by **mutations** (see page 81).

Ribosomes and rough endoplasmic reticulum

Ribosomes (figure 8.4 and Appendix 1) are small, almost spherical, structures found in all cells. Some occur freely in the cytoplasm, others are found attached to the **endoplasmic reticulum**. Ribosomes are the site of protein synthesis.

The rough endoplasmic reticulum (rough ER) is illustrated in figure 8.4 and Appendix 1. It is composed of a system of flattened sacs and tubules and bears ribosomes on the outer surface of its membranes. Rough ER is continuous with the outer nuclear membrane and provides a large surface area for chemical reactions to occur. It is present throughout the cell and acts as a pathway for the transport of materials (e.g. protein made in the ribosomes) through the cell. Rough ER is especially abundant in growing cells where much protein synthesis goes on.

Secretion of proteins

The **Golgi apparatus** (figure 8.5 and Appendix 1) is composed of a group of flattened fluid-filled sacs formed from vesicles pinched off the rough ER, containing newly synthesised protein.

The Golgi apparatus *processes* this protein (e.g. by adding a carbohydrate part to it), making it ready to leave the cell, as **glycoprotein**. Small fluid-filled sacs, called **vesicles**, containing the

finished product (e.g. digestive enzyme, hormone etc.) become pinched off the ends of the Golgi apparatus.

Figure 8.5 shows how these vesicles move towards and fuse with the cell membrane. The contents of a vesicle are then discharged to the outside. This release of intracellular products from the cell is called **secretion**.

The Golgi apparatus is especially active in secretory cells such as gastric gland cells in the stomach wall (which produce the enzyme pepsin) and certain cells in the pancreas (which produce the hormone insulin).

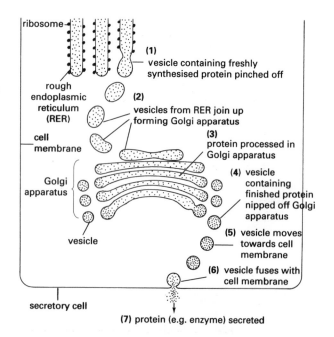

Figure 8.5 *Processing and secretion of a protein*

QUESTIONS

1 Draw a diagram of the mRNA strand that would be transcribed from section **X** of the DNA molecule shown in Figure 8.6.

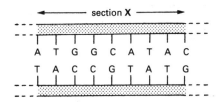

Figure 8.6

2 Name THREE ways in which RNA and DNA differ in structure and chemical composition.

3 Using the information given in table 8.1:

amino acid	base anticodon (tRNA)
asparagine	UUA
glutamic acid	CUU
proline	GGA
threonine	UGG
tyrosine	AUA

Table 8.1

a) write down the base codon (mRNA) for each amino acid;
b) draw a diagram of the sequence of amino acids that would be formed from the portion of mRNA shown in Figure 8.7

A A U G A A U A U G A A C C U A A U A C C
↑ start here

Figure 8.7

4 Copy and complete table 8.2.

stage of synthesis	site
formation of mRNA	
collection of amino acid by tRNA	
formation of codon-anticodon links	

Table 8.2

5 After protein synthesis, what finally happens to:

a) the tRNA molecule?
b) the mRNA molecule?
c) the completed protein?

6 The chemical composition of the carbohydrate and fat components of the human body are common to everyone. Explain why it is the body's **protein** component that results in inherited differences between individuals, making each person different from everyone else.

7 The information in table 8.3 refers to the relative numbers of ribosomes present in the cells of a new leaf developing at a shoot tip.

age of new leaf (in days)	relative numbers of ribosomes in cells
1	300
3	500
5	650
7	450
9	210
11	200
13	200
15	200

Table 8.3

a) Construct a hypothesis to account for the trend shown by this data.
b) Explain why the relative numbers of ribosomes in cells of a fully grown leaf do not drop to zero.

8 Figure 8.8 shows part of a secretory cell viewed under an electron microscope.

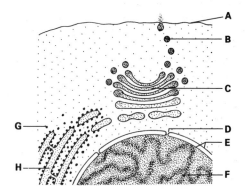

Figure 8.8

a) Name parts **A-H**.
b) Name the type of chemical that would be present in **B**.
c) State which letter indicates the site at which synthesis of this substance from its sub-units occurs.
d) Briefly explain how the sequence of these sub-units is originally determined.

9 Variety of proteins

Structure of protein

Proteins are organic compounds. In addition to always containing the elements carbon (C), hydrogen (H), oxygen (O) and **nitrogen** (N), they often contain sulphur (S) and sometimes phosphorus (P).

Each protein is built up from a large number of sub-units called **amino acids** (of which there are about twenty different types). These are joined together by **peptide bonds** to form **polypeptides** which in turn become linked together to form either **fibrous** or **globular** protein as shown in figure 9.1.

Variety of proteins

An enormous number of different proteins exist amongst living things. They vary from species to species. A human being possesses over 10 000 different proteins.

Fibrous proteins

These are insoluble in water and consist of long parallel polypeptide chains (like a rope made of strands of string). They are **structural** proteins which often have contractile properties. Examples include collagen in bone and tendons, actin and myosin in skeletal muscle, elastin in ligaments and keratin in hair.

Globular proteins

These are not truly soluble but they form a colloidal suspension in water. Their polypeptide chains are folded into a roughly spherical shape (like a tangled ball of string). They are found inside all living cells where they form a vital component of cytoplasm and play several roles.

Structural protein
Globular protein is one of the two components which make up the membrane surrounding a living cell (see page 7). It forms an essential part of all the membranes possessed by subcellular structures within the cell such as the nucleus and mitochondria. So this type of protein plays a vital **structural** role in every living cell.

Enzymes
All **enzymes** (biological catalysts) are made of globular protein. Each is folded in a particular way to expose an active surface which readily combines with a specific substrate. Since intercellular enzymes **speed up** the rate of biochemical processes such as photosynthesis, respiration and protein synthesis, they are essential for the maintenance of life.

Hormones
These are **chemical messengers** transported in an animal's blood to 'target' tissues where they exert a specific effect. Some hormones are made of globular protein and exert a **regulatory** effect on the animal's growth and metabolism.

The hormone **thyroxine**, for example, plays an essential role in the control of human growth and metabolism (see page 149). **Insulin** and **glucagon** control blood sugar level (see page 176).

Antibodies
Although Y-shaped rather than spherical, **antibodies** are also a type of globular protein. They are made by lymphocytes and **defend** the body against antigens (see page 54).

Conjugated proteins

Each of these consists of a globular protein associated with a **non-protein chemical.** **Haemoglobin** (the oxygen-carrying pigment in blood) possesses **haem**, a non-protein part which is rich in **iron. Cytochrome** (the hydrogen carrier in the cell's aerobic respiratory pathway) also contains iron.

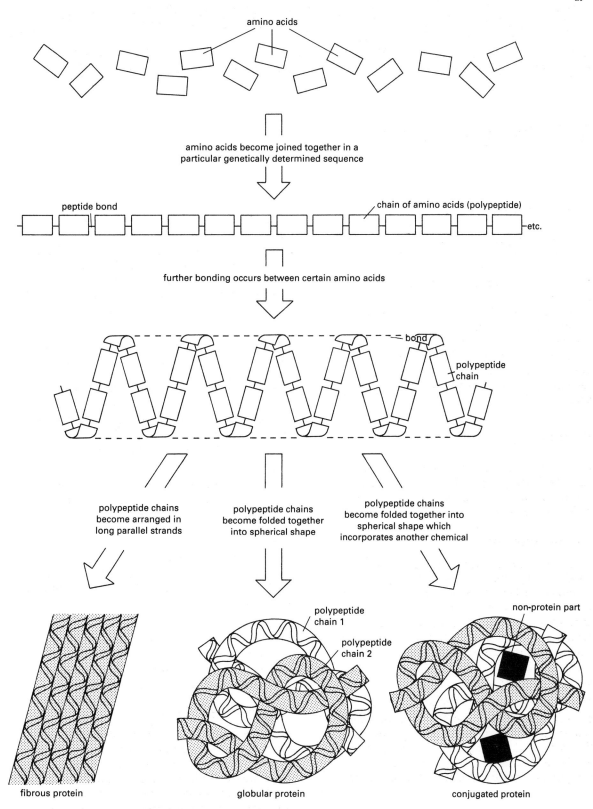

amino acids

amino acids become joined together in a
particular genetically determined sequence

peptide bond

chain of amino acids (polypeptide)

etc.

further bonding occurs between certain amino acids

bond

polypeptide
chain

polypeptide chains
become arranged in
long parallel strands

polypeptide chains
become folded together
into spherical shape

polypeptide chains
become folded together into
spherical shape which
incorporates another chemical

non-protein part

polypeptide
chain 1

polypeptide
chain 2

fibrous protein

globular protein

conjugated protein

Figure 9.1   Structure of proteins

Thus proteins play a wide variety of roles in cells. Some form part of the cell's framework, some defend the body's cells whilst others regulate biochemical reactions and metabolic processes.

QUESTIONS

1 Name the chemical element always present in protein but absent from carbohydrates.
2 Describe TWO ways in which polypeptide chains can become arranged to form a protein.
3 Name THREE types of globular protein and briefly describe the role played by each one.
4 Figure 9.2 is incomplete. It represents the action of an enzyme on its substrate. Draw a complete version of the diagram by replacing boxes **P, Q** and **R** with the omitted molecules.

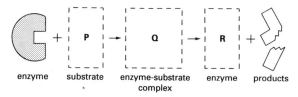

Figure 9.2

5 Some amino acids can be synthesised by the body from simple compounds. Others cannot be synthesised and must be supplied in the

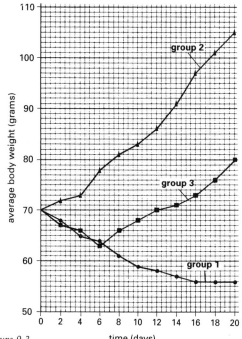

Figure 9.3

diet. The latter are called the **essential** amino acids.

The graph in Figure 9.3 shows the results of an experiment using rats: group 1 was fed zein (maize protein), group 2 was fed casein (milk protein) and group 3 was fed a varied diet.

a) One of the proteins contains all of the essential amino acids whereas the other lacks two of them. Identify each protein and explain how you arrived at your answer.

b) (i) State which protein was given to the rats in group 3 during the first six days of the experiment.

(ii) Suggest TWO different ways in which their diet could have been altered from day 6 onwards to account for the results shown in the graph.

c) Calculate the percentage decrease in average body weight shown by the rats in group 1 over the 20-day period.

d) By how many times did the average body weight of the rats in group 2 increase over the 20-day period?

DATA INTERPRETATION

People suffering from sickle cell anaemia are found to possess molecules of abnormal haemoglobin (haemoglobin S) in their red blood cells. The difference between haemoglobin S and normal haemoglobin (haemoglobin A) can be detected by the procedure shown in figure 9.4.

Once a sample of each type of haemoglobin has been digested by an enzyme, the resulting peptides are separated using the combined techniques of electrophoresis and chromatography. The characteristic chromatograms produced are identical except for the position of one peptide.

The amino acid composition of both peptide X and peptide X^1 is determined by first breaking down each peptide to its component amino acids, and then separating them by further chromatography. Subsequent sequencing analysis shows the component amino acids in each peptide to occur in the order shown in figure 9.5.

a) Sickle cell anaemia is characterised by shortness of breath, emaciation, heart failure and normally death. What basic role must haemoglobin S be failing to perform efficiently? (1)
b) Which property of peptides allows them to be separated by electrophoresis? (1)
c) (i) Describe the procedure involved in carrying

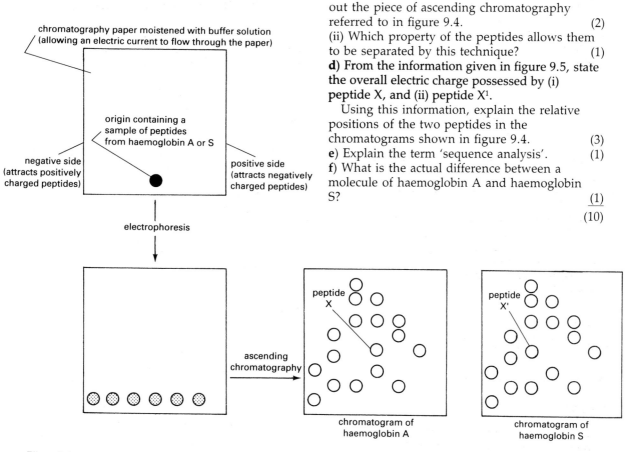

chromatography paper moistened with buffer solution
(allowing an electric current to flow through the paper)

origin containing a
sample of peptides
from haemoglobin A or S

negative side
(attracts positively
charged peptides)

positive side
(attracts negatively
charged peptides)

electrophoresis

ascending
chromatography

peptide X

chromatogram of
haemoglobin A

peptide X'

chromatogram of
haemoglobin S

Figure 9.4

out the piece of ascending chromatography referred to in figure 9.4. (2)
(ii) Which property of the peptides allows them to be separated by this technique? (1)
d) From the information given in figure 9.5, state the overall electric charge possessed by (i) peptide X, and (ii) peptide X^1.
Using this information, explain the relative positions of the two peptides in the chromatograms shown in figure 9.4. (3)
e) Explain the term 'sequence analysis'. (1)
f) What is the actual difference between a molecule of haemoglobin A and haemoglobin S? (1)
(10)

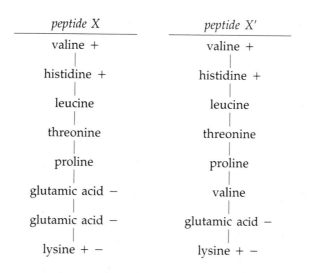

peptide X	*peptide X'*
valine +	valine +
histidine +	histidine +
leucine	leucine
threonine	threonine
proline	proline
glutamic acid −	valine
glutamic acid −	glutamic acid −
lysine + −	lysine + −

Figure 9.5

What you should know (CHAPTERS 7–9)

1 **DNA** consists of two strands twisted into a **double helix**. Each strand is made up of **nucleotides** each of which possessees one of four types of base (adenine, thymine, guanine or cytosine).

2 **Adenine** always pairs with **thymine; guanine** always pairs with **cytosine**.

3 DNA is unique because it is able to reproduce itself by **replication**. This allows the genetic message to be passed on from cell to cell, and from generation to generation.

4 RNA consists of a **single** strand of nucleotides which contain **uracil** in place of thymine and **ribose** instead of deoxyribose sugar.

5 **Messenger RNA** (mRNA) is transcribed from a strand of DNA and carries this genetic message to ribosomes in the cytoplasm. Here it meets molecules of **transfer RNA** (tRNA), each carrying an amino acid.

6 **Protein synthesis** occurs in **ribosomes**; mRNA's triplets of bases (**codons**) are 'read' and matched by tRNA's **anticodons**.

This enables **peptide bonds** to form between adjacent amino acids.

7 **Rough endoplasmic reticulum** bears ribosomes on its outer surface.

8 Freshly synthesised protein is transported via the endoplasmic reticulum to the **Golgi apparatus** where it is processed and packaged in **vesicles**.

9 Some protein is **secreted** out of the cell by vesicles moving towards, and fusing with, the plasma membrane.

10 In addition to carbon, hydrogen and oxygen, proteins always contain **nitrogen**.

11 A **protein** consists of sub-units called **amino acids** (of which there are about twenty types). Amino acids are joined together by peptide bonds to form **polypeptides**.

12 A molecule of **fibrous** protein consists of parallel polypeptide chains and has a **structural** function.

13 A molecule of **globular** protein consists of polypeptide chains folded into a **spherical** shape. Some are **structural** (e.g. those in the plasma membrane); others act as **enzymes, hormones** or **antibodies**.

10 Viruses

Viruses are micro-organisms which exhibit living and non-living characteristics. Since they can only reproduce within the living cells of another organism (the **host**), they are described as **obligate parasites**. Animals, plants and bacteria are all susceptible to invasion by viruses. Since the host cell is subsequently destroyed, viruses are always associated with **disease**. A few examples are given in table 10.1.

disease (or syndrome) caused by virus	host organism	effect
poliomyelitis	human	inflammation of spinal cord followed by paralysis
AIDS	human	destruction of immune system
foot and mouth	farm animals	development of lesions in mouth
bushy stunt	tomato plant	stunted growth
leaf mosaic	potato plant	malformation of leaves

Table 10.1   Viral diseases

Viruses are much smaller than bacteria and can only be seen with the aid of an electron microscope.

Transmission of viruses

Coughing and sneezing spread the viruses that cause respiratory infections such as pneumonia. The poliomyelitis virus is excreted in faeces and passed to food by flies. The AIDS virus is transmitted by blood and sexual contact.

Viruses which attack plants are spread by insects such as greenfly and by physical contact between leaves and between roots. Seeds, bulbs and cuttings can also carry viruses.

Invasion of a cell by a virus

A virus consists of one type of nucleic acid (either DNA or RNA) surrounded by a protective coat of protein. Its genetic material carries all the information necessary for viral multiplication.

The virus shown in figure 10.1, attacks bacteria and is called a **bacteriophage**. Its DNA thread is contained inside a head from which a tail projects. Figure 10.2 shows the invasion of a bacterium (the host) by a bacteriophage.

Alteration of cell instructions

Although completely inactive outside the host cell, once inside, the virus assumes control of the cell's biochemical machinery. It depends on the host cell for energy (from ATP) and a supply of nucleotides and amino acids to build new viral particles.

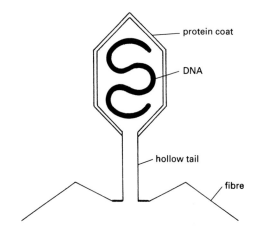

Figure 10.1   Structure of bacteriophage

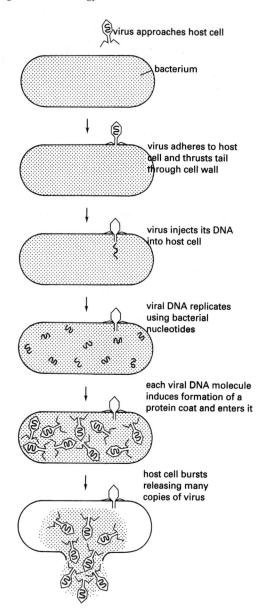

Figure 10.2 *Multiplication of a virus*

The viral DNA (or RNA) first suppresses the cell's normal nucleic acid replication and protein synthesis. Then it uses the host's nucleotides and amino acids to produce many **identical copies** of the viral nucleic acid and an appropriate number of protein coats.

In some cases the whole virus enters the host cell but the final outcome is the same: release of many copies of the virus capable of infecting new host cells and causing disease.

Some viruses infect a cell and remain dormant for an indefinite period before undergoing reproduction and release.

QUESTIONS

1 State TWO structural features possessed by all viruses.
2 Under what circumstances is a virus (i) active, (ii) inactive?
3 The following steps occur during multiplication of a bacteriophage. Arrange them into the correct order.
   A assembly of completed viruses;
   B alteration of host cell's biochemistry;
   C attachment of virus to host cell;
   D production of protein coats;
   E release of new virus particles;
   F injection of DNA into host cell;
   G replication of viral DNA.
4 In the absence of a suitable host cell, a virus exhibits none of the characteristics of living things and may adopt a crystalline form.
   a) Why then do scientists consider viruses to be living things?
   b) Scientists do not, however, classify a virus particle as a cell. Suggest why not.
5 The protein coat that surrounds the nucleic acid molecule of a virus is called a **capsid**. It consists of sub-units (capsomers) assembled into either a helical (spiral) or a polyhedral (many-sided) formation. Some capsids are in

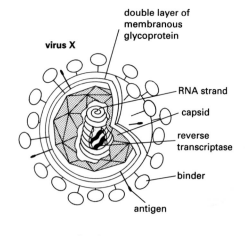

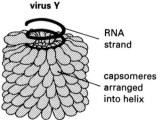

Figure 10.3

type of nucleic acid	shape of capsid	state of capsid	example	host cell of example
RNA	helical	naked	tobacco mosaic virus	plant
		enveloped	influenza virus	animal
	polyhedral	naked	bushy stunt virus	plant
		enveloped	human immunodeficiency virus	animal
DNA	helical	naked	coliphage virus	bacterium
		enveloped	smallpox virus	animal
	polyhedral	naked	bacteriophage virus	bacterium
		enveloped	herpes	animal

Table 10.2

turn surrounded by a membranous envelope.

Table 10.2 refers to eight different viruses.

a) Name THREE structural features possessed by the virus that causes smallpox.

b) (i) Identify viruses **X** and **Y** shown in figure 10.3.

(ii) Which labelled component of virus **X** is responsible for making the host cell copy the viral RNA into viral DNA before multiplication can begin?

c) Name the virus whose DNA strand is protected by an enveloped capsid composed of capsomers laid down in a polyhedral arrangement.

d) Convert table 10.2 into a key of paired statements.

11 Cellular response in defence

Response of individual cells

Interferon

On being invaded by a virus, many types of animal cell respond by producing one of the **interferons** (a family of proteins). An interferon itself does not have antiviral properties. However when it is released from the infected cell, it becomes attached to the plasma membrane of neighbouring cells and somehow activates the genes in these cells that code for the synthesis of **antiviral proteins**. These are enzymes which defend the cells by blocking the synthesis of proteins needed by the invading virus for its multiplication. This sequence of events is summarised in figure 11.1.

Although interferon works best on the cells of the species that produced it, it is not specific to one virus. Thus a few molecules of interferon can trigger the synthesis of many antiviral proteins giving effective defence against a *wide range* of viruses. By this means, viral infections can often be contained until the specific immune responses (which are slower to act) can take over (see page 54).

In addition to viruses, other pathogenic microbes with an intracellular phase in their life cycle are now known to induce cells to make interferon in response to invasion. Future mass production of human interferon by genetic engineering may provide the means by which certain diseases and forms of cancer can be treated.

Whole organism defence mechanisms

The warm, moist, food-rich conditions that exist inside the human body are ideal for the growth and multiplication of many types of disease-causing (pathogenic) micro-organisms. However these are normally prevented from causing disease by the body's **defence mechanisms**.

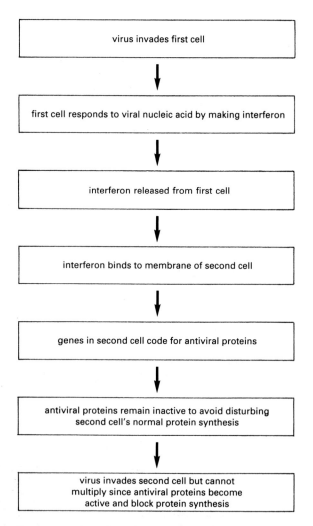

Figure 11.1   Role of interferon

First lines of defence

The entry of many microbes is prevented by the **skin** which acts as an almost impenetrable barrier. In addition **mucus** and **cilia** trap and sweep dirt and micro-organisms out of the respiratory system. **Coughing, sneezing** and **vomiting** remove unwanted foreign particles; **acid** in the stomach kills many bacteria; **ear wax** traps germs; **lysozyme**, a powerful digestive enzyme in tears and saliva, destroys bacteria; and **clotting of blood** at wounds helps to prevent microbial entry.

Second lines of defence

The body employs further methods of defence to deal with those micro-organisms that manage to bypass the first lines of defence (e.g. via an insect bite or a dirty wound).

These second lines of defence are **active mechanisms** mainly brought about by **white blood cells** which recognise and respond to the presence of invading foreign particles. Three types of white blood cell are shown in figure 11.2.

Phagocytosis

Phagocytosis ('cell-eating') is the process by which foreign bodies such as bacteria are **engulfed** and destroyed. Cells capable of phagocytosis (e.g. monocytes and neutrophils) are called **phagocytes**.

A phagocyte detects chemicals released by a bacterium and moves along a concentration gradient towards it as shown in figure 11.3. The phagocyte adheres to the bacterium and engulfs it in a vacuole formed by an infolding of the cell membrane.

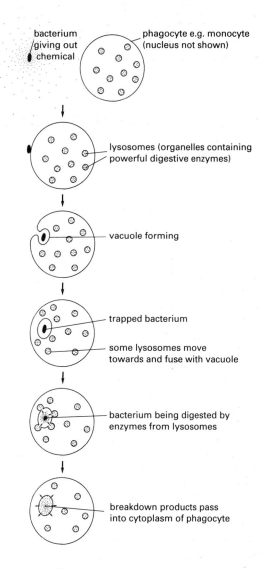

Figure 11.3   Phagocytosis

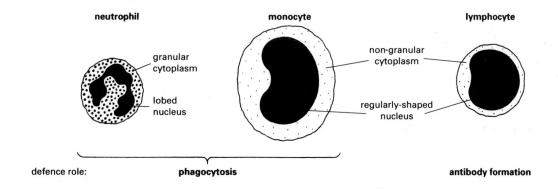

Figure 11.2   White blood cells

A phagocyte's cytoplasm contains a rich supply of organelles called **lysosomes** which contain powerful **digestive enzymes**. Some of these lysosomes fuse with the vacuole and release their enzymes into it. The bacterium becomes digested and the breakdown products are absorbed by the phagocyte.

During infection hundreds of phagocytes migrate to the infected area and engulf many bacteria by phagocytosis. Dead bacteria and phagocytes often gather at a site of injury forming **pus**.

The immune response

Immunity is an organism's ability to resist infectious disease. Phagocytosis and the action of interferon and lysozyme are examples of **non-specific** immune responses since they provide general protection against a wide range of microbial invaders.

Specific immune response (antibody formation)

An **antigen** is a complex molecule such as a protein or complex carbohydrate which is recognised as alien by the body's lymphocytes. The antigen's presence in the bloodstream stimulates lymphocytes to produce special protein molecules called **antibodies** which are specific to that antigen. The body possesses thousands of different types of lymphocytes. Each is capable of recognising and responding to *one* antigen.

A virus particle is surrounded by a coat of protein which acts as an antigen at several sites on the surface of the virus.

An antibody is a Y-shaped molecule as shown in figure 11.4. Each of its arms bears a **receptor** (binding) site whose shape is **specific** to a particular antigen. When an antibody meets its complementary antigen, the two combine at their specific sites like a lock and key and the antigen is rendered harmless (see figure 11.5).

B cells and T cells

The antibodies shown in figure 11.5 are produced and released as *free* molecules by B lymphocytes (**B cells**). A second group called T lymphocytes (**killer T cells**) form antibodies on their surfaces. These antibodies are not released but remain *cell-bound* and play their defensive role against appropriate antigens when the T cells arrive at the site of infection.

Following the destruction of antigens, some B cells and T cells remain in the body as **memory cells**, allowing a much more rapid and vigorous response to any further invasion by a previously encountered antigen.

Types of specific immunity

Active immunity

The organism produces its own antibodies in one of two possible ways.

Naturally acquired immunity
The person is exposed to the disease-causing antigen, suffers the disease, makes antibodies and continues to be able to do so for a long time (or even permanently) after recovery. The person is therefore **immune** to future attacks by that antigen.

Artificially acquired immunity (immunisation)
The person receives a small amount of vaccine containing antigens which have been treated in a certain way so that they induce antibody formation but do not cause the disease.

Passive immunity

Instead of the individual making antibodies in response to the antigen, *ready-made* antibodies are passed into his/her body. Such passive immunity occurs naturally when antibodies cross the placenta from mother to foetus (and from mother's milk to suckling baby), giving the child protection for a short time until its own immune system develops.

Passive immunity is brought about artificially by extracting antibodies that have been made by one mammal (e.g. horse) and injecting them into another (e.g. human). The effect is short-lived since the antibodies only persist for a short time.

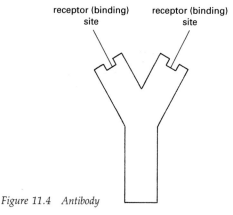

receptor (binding) site   receptor (binding) site

Figure 11.4 Antibody

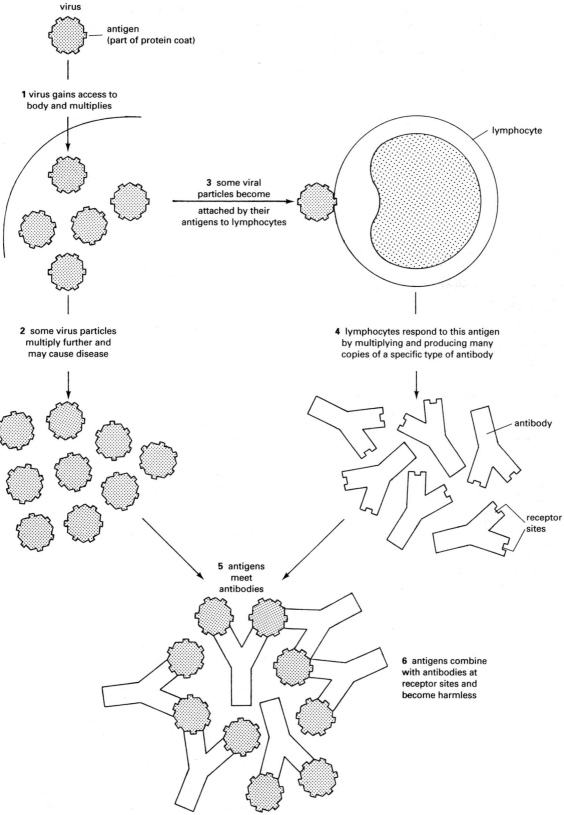

virus

antigen
(part of protein coat)

1 virus gains access to
body and multiplies

lymphocyte

3 some viral
particles become

attached by their
antigens to lymphocytes

2 some virus particles
multiply further and
may cause disease

4 lymphocytes respond to this antigen
by multiplying and producing many
copies of a specific type of antibody

antibody

receptor
sites

5 antigens
meet
antibodies

6 antigens combine
with antibodies at
receptor sites and
become harmless

Figure 11.5   Action of antibodies

However they give the person protection until his/her own immune system has time to respond to the antigen.

Rejection of transplanted tissues

When living tissue is transplanted from one individual to another, the T lymphocytes of the recipient's immune system regard the new tissue as a collection of foreign antigens and attempt to destroy them. This destruction process is called **tissue rejection** and always occurs unless donor and recipient are genetically identical twins.

Successful transplants of tissues and organs (e.g. liver, kidney etc.) are made possible by choosing a donor and a recipient that are as genetically similar as possible and then administering **immunosuppressor drugs**. Unfortunately these greatly inhibit the recipient's immune system and make the person susceptible to serious diseases such as pneumonia.

Drugs are now being developed which tend to eliminate the graft rejection response while leaving the immune response intact. In addition, scientists are working on agents which will induce immunological tolerance in advance of the tissue transplant.

Response of plant cells to invasion

Plants are constantly exposed to an immense array of pathogenic micro-organisms yet the vast majority of healthy plants manage to resist almost all of them.

The cell wall acts as a natural barrier against bacteria and viruses which normally only gain access via the biting or piercing mouthparts of an insect (e.g. aphid). Fungal pathogens however can penetrate the cell wall without insect assistance by secreting digestive enzymes.

Plant cells respond to pathogenic attack by mobilising a variety of defence mechanisms aimed at weakening and destroying the invader or strengthening the barrier against it.

Chemical responses

Phytoalexins
These include a wide range of antifungal chemicals made by plant cells in in response to fungal invasion. The production of a suitable **phytoalexin** is thought to occur as a result of the sequence of events shown in figure 11.6.

Although the plant cells in the front line of defence may show browning and even die, nearby cells are also stimulated to make phytoalexin. This creates an **antifungal environment** which prevents the invader from spreading beyond the localised site of attack.

Plants lacking the appropriate gene(s) to code for the necessary phytoalexin-synthesising enzymes are susceptible to the disease.

Hydrogen peroxide
For many years, phytoalexins were thought to be the first products formed by a plant in response to invasion. However they are not detected in the plant until 2–4 hours after the attack begins. Recent research shows that, within a few minutes, plant cells respond to the presence of an alien fungal chemical by producing **hydrogen peroxide** (H_2O_2). This is then acted upon by the enzyme peroxidase (already present in the cell wall) to oxidise and destroy the alien chemical (A) as follows:

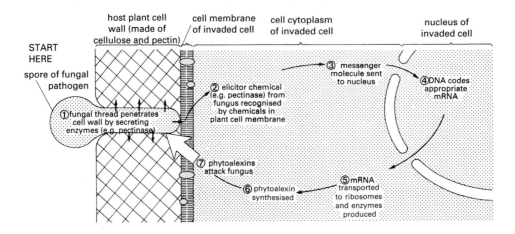

Figure 11.6   Phytoalexin production

$$AH_2 + H_2O_2 \xrightarrow{\text{peroxidase}} 2H_2O + A \text{ (oxidised)}$$

This speedy burst of **oxidative activity** is thought to be the plant cell's first real line of defence against an invading pathogen (see figure 11.7). In addition, H_2O_2 is thought to play an important role in stimulating the production of phytoalexins.

Barriers

Lignification

Lignin is a complex molecule of variable chemical composition. It is a common constituent of cell walls and the principal component of wood. Increased lignification is found to occur in the walls of many plant cells about 12–24 hours after fungal attack.

The extra lignin protects the cell against further invasion by acting as a *barrier* which is resistant both to mechanical penetration by fungal threads and degradation by fungal enzymes.

Callose

Callose is a complex polysaccharide of glucose. Following injury, sieve tubes in many plants are found to become sealed off by heavy callose deposits. These are thought to prevent the invading microbe from spreading through the plant via the phloem tissue.

Ethylene

Ethylene is a gas released by most plant organs (especially ripening fruit). Increased ethylene production is found to occur in the leaves of various plants following infection by viruses. The extra ethylene accelerates the **abscission** (natural cutting off and separation) of damaged leaves, thereby ridding the plant of the diseased organ.

Galls and tannins

When a parasite such as an insect or fungus successfully penetrates the tissues of a host plant, the latter often produces a **gall** in response to a chemical stimulus made by the parasite.

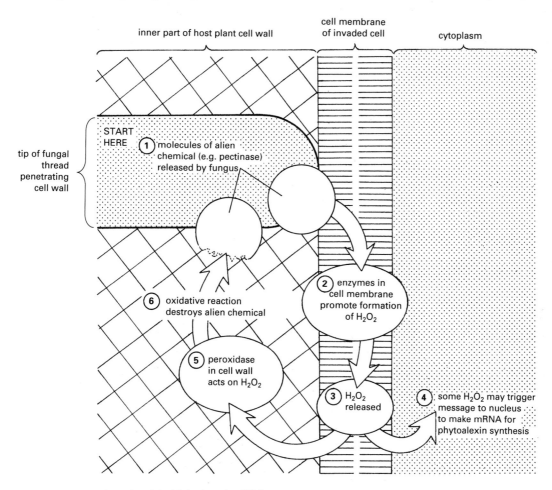

Figure 11.7 Hypothetical model of defensive role of H_2O_2

A gall is an abnormal swelling of plant tissues resulting from active division of the cells at the site of injury. Many galls contain acidic chemicals called **tannins** which are thought to play a protective role.

This combination of extra layers of cells and rich deposits of tannin in a gall provides the plant with a *protective barrier* within which such a parasite (and its toxins) can be isolated (see figure 11.8) and prevented from causing further damage.

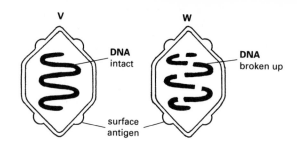

Figure 11.9

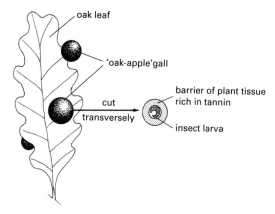

Figure 11.8   Gall

QUESTIONS

1 Classify the following processes that occur in the human body into first and second lines of defence:
blood clotting, antibody production, phagocytosis, lysozyme production.

2 Arrange the following steps into the correct sequence in which they occur during phagocytosis:
a) bacterium digested by enzymes;
b) bacterium engulfed;
c) lysosomes fuse with vacuole;
d) phagocyte meets bacterium.

3 Suggest how a monocyte avoids self-digestion during the process of phagocytosis.

4 a) Which type of blood cell produces antibodies?
b) Using a diagram to illustrate your answer, explain what is meant by the specificity between an antigen and an antibody.

5 Figure 11.9 shows a disease-causing virus (**V**) and an attenuated (weakened) version of it (**W**).

a) Suggest why receiving a vaccine containing **W** gives immunity against attacks in the future by **V**.
b) Would such immunity be described as naturally or artificially acquired?

6 a) Why must immunosuppressor drugs normally be administered to patients undergoing kidney transplant surgery?
b) What problem may result from the use of these drugs?

7 Briefly describe a method adopted by a plant to counteract invasion by a fungus where the latter is weakened or destroyed.

8 A certain type of insect gall on willow trees only develops if the parasite's egg hatches into a larva. Eventually, when the larva stops feeding and becomes a pupa, gall development stops.
a) From this relationship, identify the gall-maker.
b) (i) Suggest which stage in the life cycle of the gall-causer produces the chemical stimulus to which the gall-maker responds.
(ii) Briefly describe how your suggestion could be tested experimentally.
c) What benefit is gained by a gall-maker in return for the materials and energy expended during gall formation?

DATA INTERPRETATION

Table 11.1 refers to antibody proteins, called immunoglobulins, found in human blood.
The graph in figure 11.10 refers to the sequence of events which occurs in response to two separate injections of a type of antigen into a small mammal.
a) (i) Which immunoglobulin in the table would be found in the blood of an unborn baby?
(ii) Suggest why these antibodies are only needed

	immunoglobulin (Ig)				
	IgA	IgD	IgE	IgG	IgM
molecular weight	170 000	184 000	188 100	150 000	960 000
normal serum concentration	1.4–4.0 g/l	0.1–0.4 g/l	0.1–1.3 mg/l	8.0–16.0 g/l	0.5–2.0 g/l
ability to cross placenta	no	no	no	yes	no

Table 11.1

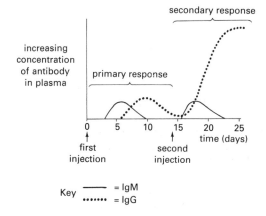

Figure 11.10

by the baby for a few months after birth. (2)
b) Of the five types of immunoglobulin molecules, which is the (i) largest, (ii) rarest? (2)
c) State the normal serum concentration of IgA in mg/ml. (1)
d) With reference to IgM and IgG, state ONE feature common to both the primary and secondary response shown in figure 11.10 (1)
e) With reference to IgG, state THREE differences between the primary and the secondary response. (3)
f) Antibodies such as IgG are now known to be produced by the activity of long-lived lymphocytes. With reference to the graph, suggest why the latter are called 'memory cells'. (1)
(10)

What you should know (CHAPTERS 10–11)

1 **Viruses** exhibit living and non-living characteristics and can only reproduce within the living cells of another organism.
2 A virus consists of one type of **nucleic acid** surrounded by a coat of **protein**.
3 Once inside a host cell, the virus **alters** the host cell's instructions to produce many identical copies of itself.
4 Some individual cells respond to viral attack by producing **interferon** which stimulates neighbouring cells to produce **antiviral proteins**.
5 A vertebrate organism possesses first and second **lines of defence** against micro-organisms.
6 Second lines of defence depend mainly upon the activities of **white blood cells**.
7 **Monocytes** engulf and destroy microbes by **phagocytosis**. This is a non-specific immune response.
8 **Lymphocytes** recognise **antigens** on the surface of a microbe and produce **antibodies**. These possess **receptor sites** which bind to one particular type of antigen rendering it harmless. This is a **specific** immune response.
9 Plants do not make antibodies. They defend themselves against invasion by producing **chemicals** (enzymes and phytoalexins)
10 Plants also employ physical defences by forming **galls** and other **barriers**.

12 Meiosis

Role of sexual reproduction

Continuous and discontinuous **variation** exist amongst the members of a species. Much of this variation is determined by genes and is **inherited**. For a species to survive, reproduction must occur. During **sexual** reproduction, the nuclei of two sex cells (**gametes**) fuse to form a **zygote**. Since the two sex cells are genetically different from one another, the zygote formed is different from either of the parents. Thus sexual reproduction maintains and increases genetic variation within a population.

In the long term, such variety is of great importance because it helps a species to adapt to a changing environment. Imagine, for example, that a new disease appears. If great genetic variation exists amongst the members of the species, then there is a good chance that some of them will be resistant and survive. If they were all identical and susceptible to the disease, the whole species would be wiped out. Genetic variation may also allow some members of a species to colonise new environments.

This chapter explores the means by which the gametes in a breeding population receive different **alleles** of genes and thereby provide the raw material for evolution to occur.

Haploid and diploid cells

The nucleus of a cell contains a complement of **chromosomes** which varies in number from species to species. A **haploid** cell (e.g. a gamete) has a **single** set of chromosomes (i.e. one of each type) whereas a **diploid** cell (e.g. a zygote) has a **double** set of chromosomes (i.e. two of each type which form homologous pairs).

A single chromosome complement is represented as **n**. Therefore a haploid gamete has a **ploidy number n** and a diploid zygote **2n**. In humans $n = 23$, whereas in the example shown in figure 12.1, $n = 2$.

Since chromosomes contain a cell's DNA (see page 37), the mass of DNA present in a haploid cell (e.g. sperm) is half of that found in a diploid body cell (e.g. liver). Human red blood cells lack nuclei and have neither chromosomes nor DNA.

Meiosis

Meiosis is a form of nuclear division which results in the production of four haploid (n) gametes from one diploid (2n) gamete mother cell. It occurs at specific sites in living organisms (see table 12.1).

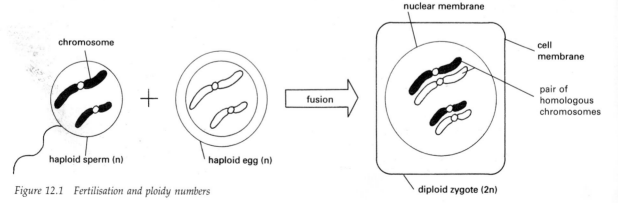

Figure 12.1 Fertilisation and ploidy numbers

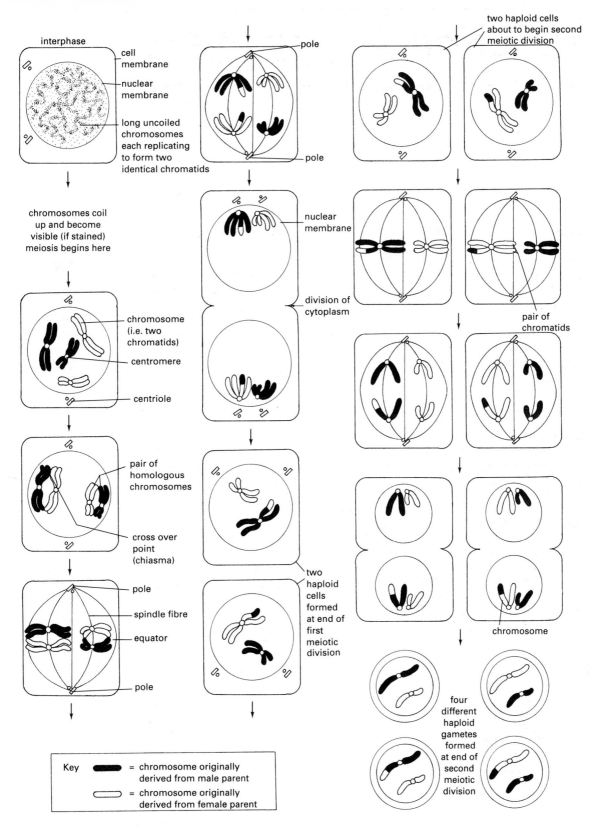

Figure 12.2   Meiosis

site of meiosis	diploid gamete mother cell	haploid gametes formed
testis of animal	sperm mother cell	sperm
ovary of animal	egg mother cell	eggs (ova)
anther of flowering plant	pollen mother cell	pollen
ovary of flowering plant	egg mother cell	eggs (ovules)

Table 12.1   Sites of meiosis

Mechanism of meiosis

Meiosis (reduction division) involves two consecutive cell divisions. The gamete mother cell divides into two and then the products divide again. Figure 12.2 refers to a gamete mother cell containing four chromosomes.

During interphase each chromosome duplicates forming two identical **chromatids** (see also page 38). Therefore when the nuclear material condenses and becomes visible (on staining) at the start of the first meiotic division,

each chromosome is seen to consist of two chromatids attached at a centromere.

Homologous chromosomes pair up and, although the chromatids tend to repel one another, homologous chromosomes remain joined at certain points called **chiasmata**. It is here that exchange of genetic material may occur by two chromatids 'swapping' portions. This is called **crossing over** (see also page 63). The nuclear membrane disappears, spindle fibres form and homologous pairs become arranged on the equator of the cell. The arrangement of each pair relative to any other is random (see also page 63).

On contraction of the spindle fibres, one chromosome is pulled to one pole and its homologous partner to the opposite pole. A nuclear membrane forms round each group of chromosomes and the cytoplasm divides.

Each of the haploid cells now undergoes the second meiotic division in the same way except that single chromosomes (each made of two chromatids) line up at each equator. On separation from its partner, each chromatid becomes a chromosome.

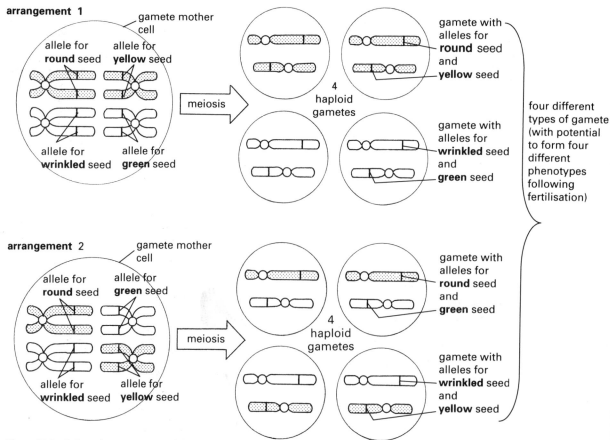

Figure 12.3   Independent assortment of chromosomes

Significance of meiosis

Prior to the mixing of one individual's genotype with that of another at fertilisation, meiosis provides the opportunity for *new combinations* of the existing genetic variation to arise as follows.

Independent assortment of chromosomes

When homologous pairs of chromosomes line up on the equator during the first meiotic division, the final position of any one pair is at random relative to another. Figure 12.3 shows a gamete mother cell in a pea plant (where only two homologous pairs have been drawn). There are two ways in which the pairs can become arranged. Subsequent meiotic divisions bring about independent assortment of chromosomes giving rise to 2^2 (i.e. 4) different genetic combinations.

A human gamete mother cell with 23 homologous pairs has the potential to produce 2^{23} (i.e. 8388608) combinations.

Crossing over between homologous chromosomes

When portions of genetic material are exchanged as shown in figure 12.4 of a fruit fly cell (where only one homologous pair has been drawn), four different types of gamete arise and variation is further increased.

Comparison of mitosis and meiosis

These two processes of nuclear division (followed by cell division) are compared in table 12.2.

	mitosis	meiosis
site of division	occurs in flowering plant's meristems and all over body of growing animal	occurs only in gamete mother cells in sex organs
pairing and movement of chromosomes	no pairing of homologous chromosomes; chromosomes line up singly on equator	homologous chromosomes form pairs; chromosomes line up in pairs on equator
exchange of genetic material	chiasmata not formed and no crossing over occurs	chiasmata may be formed and crossing over may occur
outline of division	single division of nucleus	double division of nucleus
number and type of cells produced	following cell division, two identical daughter cells formed	following cell division, four genetically different gametes formed
effect on chromosome number	chromosome number unaltered	chromosome number halved
effect on variation	does not increase variation within a population	increases variation by providing opportunity for independent assortment and crossing over to occur

Table 12.2   Comparison of mitosis and meiosis

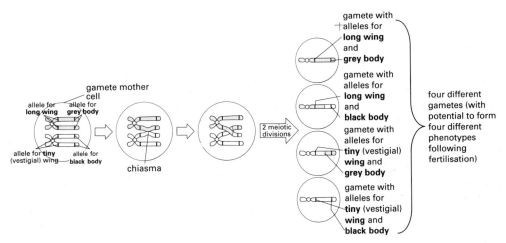

Figure 12.4   Crossing over

```
┌─────────────────────────────┐
│                             │
│         QUESTIONS           │
│                             │
└─────────────────────────────┘
```

1 Explain the difference between the terms haploid and diploid.
2 Compare the processes of mitosis and meiosis by writing one or two sentences about each difference. (Do NOT draw up a table.)
3 With reference to the cell in Figure 12.5 state the number of:

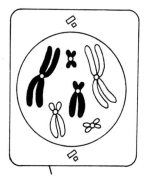

**egg mother cell
at start of meiosis**

Figure 12.5

a) chromatids present;
b) centromeres present;
c) chromosomes present;
d) pairs of homologous chromosomes present;
e) chromosomes that would be present in each gamete produced.
f) (i) Draw diagrams to show TWO different ways in which the homologous pairs of chromosomes could become arranged at the equator during the first meiotic division.
(ii) Give the gametes that would result from each arrangement (if no crossing over occurred).
4 Figure 12.6 shows a pair of homologous chromosomes during meiosis.

Figure 12.6

a) How many chiasmata occur between chromatids X and Y?
b) Draw a diagram showing the appearance of these chromosomes at the end of the first meiotic division.

13 Monohybrid cross

A cross between two true-breeding parents which differ by **one** respect is called a **monohybrid cross**.

Gregor Mendel (1822–84), an Austrian monk, performed early monohybrid crosses using varieties of pea plant which possessed characteristics showing discontinuous variation. By appreciating the importance of working with **large numbers** of plants, studying **one characteristic** at a time and **counting** the offspring produced, Mendel was the first to put genetics on a firm scientific basis.

In the experiment shown in figure 13.1, Mendel crossed pea plants which were true-breeding for production of round seeds with pea plants true-breeding for wrinkled seeds.

All the seeds produced in the **first filial generation (F_1)** were round. Once these seeds had grown into plants, they were self-pollinated. The resultant **second filial generation (F_2)** consisted of 7324 seeds (5474 round and 1850 wrinkled). This is a ratio of 2.96:1.

When these figures are analysed statistically (see Appendix 5) they are found to represent a ratio of 3 round : 1 wrinkled. An exact 3:1 ratio is rarely obtained because pollination and fertilisation are **random** processes both of which involve an element of chance.

The difference between the expected ratio and the observed ratio is known as the **sampling error**. The larger the sample, the smaller the error.

Since wrinkled seed, absent in the F_1, reappears in the F_2, 'something' has been transmitted undetected in the gametes from generation to generation. Mendel called this a **factor**. Today we call it a **gene**. In this case it is the gene for seed shape which has two forms (**alleles**) round and wrinkled.

Since the presence of the round allele masks the presence of the wrinkled allele, round is said to be **dominant** and wrinkled **recessive**.

Using symbols 'R' for round and 'r' for wrinkled, the cross can be summarised as follows:

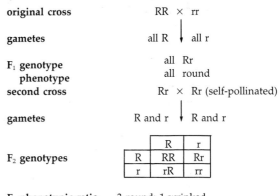

original cross		RR × rr	
gametes		all R ↓ all r	
F_1 genotype		all Rr	
phenotype		all round	
second cross		Rr × Rr (self-pollinated)	
gametes		R and r ↓ R and r	

		R	r
F_2 genotypes	R	RR	Rr
	r	rR	rr

F_2 phenotypic ratio   3 round: 1 wrinked

Mendel's first law

From the results of his many monohybrid crosses, Mendel formulated his first law, the **principle of segregation**. Expressed in modern terms it states that: *the alleles of a gene exist in pairs but when gametes are formed, the members of each pair pass into different gametes. Thus each gamete contains only one allele of each gene.*

Incomplete dominance

In some cases it is found that one allele of a gene is not completely dominant over the other allele. For example when true-breeding red snapdragon plants are crossed with true-breeding white (ivory) plants, the F_1 are all pink. When the F_1 is self-pollinated, the F_2 occurs in the ratio 1 homozygous red:2 heterozygous pink:1 homozygous ivory. This cross is shown in figure 13.2 where R=allele for red flower and I=allele for ivory flower.

parent plant true-breeding
for round seeds

x

parent plant true-breeding
for wrinkled seeds

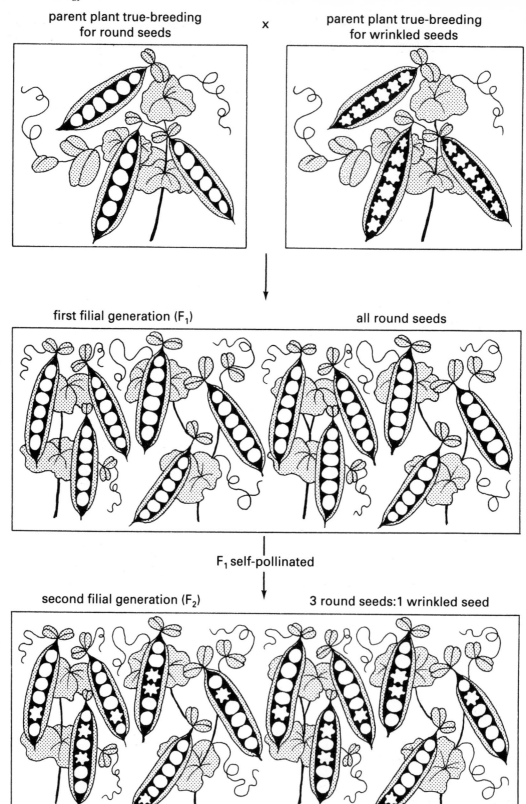

first filial generation (F₁)

all round seeds

F₁ self-pollinated

second filial generation (F₂)

3 round seeds:1 wrinkled seed

Figure 13.1   Monohybrid cross

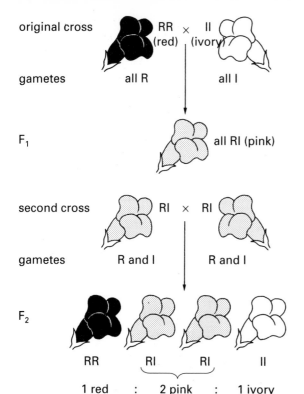

genotype(s)	phenotype (coat colour)
CC^{ch}, CC^H, CC^a	brown
$C^{ch}C^H$, $C^{ch}C^a$	grey
C^HC^a	black and white

Table 13.1   Multiple alleles for coat colour

genotype(s)	phenotype (blood group)
AA, AO	A
BB, BO	B
AB	AB
OO	O

Table 13.2   Multiple alleles for blood group

Figure 13.2   Incomplete dominance

In all cases of incomplete dominance, heterozygous individuals are phenotypically different from both of their homozygous parents.

Multiple alleles

Mendel never found more than two alleles for any one gene. However in some cases several alleles of a gene exist and are available to occupy that particular gene site (**locus**) on a chromosome. Obviously each diploid individual can only possess a maximum of two of these alleles (one from the male parent and one from the female parent).

Within a population of domestic rabbits, four alleles of the gene for coat colour exist. These show complete dominance in the order C (brown) dominant to C^{ch} (chinchilla) dominant to C^H (Himalayan) dominant to C^a (albino).

The phenotype of each homozygous genotype is shown in figure 13.3. Table 13.1 gives the phenotypes of the heterozygous genotypes.

Similarly the gene for blood group in humans has three alleles. A and B are co-dominant to recessive O (see table 13.2).

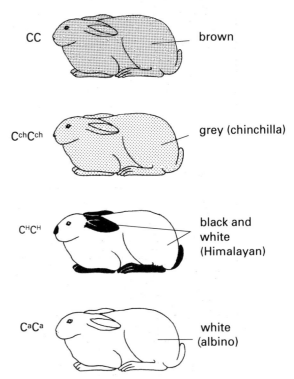

Figure 13.3   Multiple alleles for coat colour

QUESTIONS

1 In guinea pigs, long hair is recessive to short hair. A long-haired female was crossed with a heterozygous male.
a) Using symbols of your choice, present this cross as a diagram.
b) What proportion of the offspring will be long-haired?

2 Figure 13.4 shows a human family tree. The ability to roll the tongue is governed by the presence of the dominant allele R. The recessive allele is r.

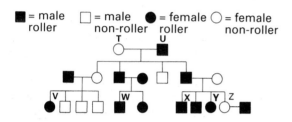

= male roller   □ = male non-roller   ● = female roller   ○ = female non-roller

Figure 13.4

a) How many male grandchildren do grandparents **T** and **U** have?
b) From the choice given in brackets, identify the person who could have the genotype RR. (**V, W, X, Y**).

c) Person **Z** marries a person of genotype Rr. The chance of each of their children being a tongue roller is:
(i) 1 in 1   (ii) 1 in 2   (iii) 1 in 3   (iv) 1 in 4

3 In rabbits, spotted coat is dominant to uniform coat. Show in a diagram the cross that should be set up to find out if a rabbit with a spotted coat is homozygous or heterozygous for this gene.

4 In shorthorn cattle, red coat colour is incompletely dominant to white coat. Heterozygotes which possess a mixture of red and white hairs are described as 'roan'.
a) Using suitable symbols, show in diagrammatic form the outcome of each of the following crosses:
(i) red × red, (ii) roan × roan, (iii) roan × white, (iv) red × white.
b) In each case, state the phenotypes of the offspring and the ratio in which they occur.

5 In mice, the gene that determines coat colour has four alleles, homozygous combinations of which produce the phenotypes: CC = full colour, $C^{ch}C^{ch}$ = chinchilla, $C^{H}C^{H}$ = Himalayan, and $C^{a}C^{a}$ = albino.

The alleles show complete dominance in the order: $C > C^{ch} > C^{H} > C^{a}$. Table 13.3 shows the outcome of two crosses. Work out the genotypes of the parents in each cross.

cross	phenotypes of parents	phenotypes of offspring			
		full colour	chinchilla	Himalayan	albino
1	full colour × albino	4	0	5	0
2	chinchilla × Himalayan	0	6	3	3

Table 13.3

14 Dihybrid cross

A cross between two true-breeding parents, which differ from one another by *two* respects, is called a **dihybrid cross**. In one of his experiments, Mendel crossed true-breeding pea plants which produced round yellow seeds with true-breeding plants which produced wrinkled green seeds.

This cross involves two genes: the gene for seed shape (alleles – round and wrinkled) and the gene for seed colour (alleles – yellow and green).

All the F_1 plants produced round yellow seeds. This shows that round is dominant to wrinkled,

Let R = allele for round

and r = allele for wrinkled

Let Y = allele for yellow

and y = allele for green

original cross	RRYY × rryy
gametes	all RY ↓ all ry
F_1	all Rr Yy
second cross	RrYy × RrYy (F_1 self-fertilised)
gametes	RY, Ry ,rY, ry ↓ RY ,Ry, rY, ry

gametes

♀ \ ♂	RY	Ry	rY	ry
RY	RRYY	RRYy	RrYY	RrYy
Ry	RRYy	RRyy	RrYy	Rryy
rY	rRYY	rRYy	rrYY	rrYY
ry	rRyY	rRyy	rryY	rryy

gametes

F_2 (phenotypic ratio) = 9 round yellow ◯ :3 round green ◓

3 wrinkled yellow ✽ :1 wrinkled green ✾

Figure 14.2 Dihybrid cross using symbols

and yellow is dominant to green. This cross was followed through to the F_2 generation. A simplified version is shown in figure 14.1. The actual observed F_2 consisted of 315 round yellow seeds, 108 round green, 101 wrinkled yellow and 32 wrinkled green.

When these figures are analysed statistically, they are found to represent a 9:3:3:1 ratio. Figure 14.2 shows how an F_2 with such a phenotypic ratio can arise. For the sixteen possible combinations shown in the Punnett square to occur, each F_1 parent at the second cross must make four different types of gamete in equal numbers.

How does this occur? If the genes for seed shape and colour are located on different chromosomes then these will be subjected to independent assortment during meiosis as shown in figure 12.3 on page 62. There are two possible ways in which homologous pairs can become arranged at the equator. As a result, approximately 50% of gamete mother cells produce RY and ry gametes, and the other 50% produce Ry and rY gametes.

Mendel's second law

From the results of his dihybrid crosses, Mendel was the first to appreciate this idea and used it to formulate his second law, the **principle of independent assortment**. Expressed in modern terms it states that: *during gamete formation, the two alleles of a gene segregate into different gametes, independently of the segregation of the two alleles of another gene.*

So in the above cross, there is just as much chance of R and y ending up together in a gamete as R and Y going together. This principle can be demonstrated by carrying out the bead model experiment of gamete formation shown in figure 14.3.

parent plant true-breeding
for round yellow seeds   ×   parent plant true-breeding
for wrinkled green seeds

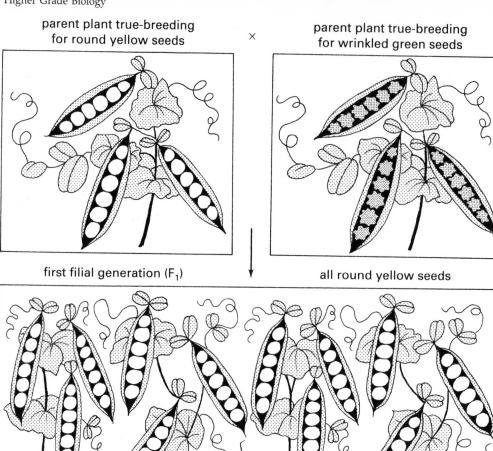

first filial generation (F₁)   all round yellow seeds

F₁ self-pollinated

second filial generation (F₂)   9 round yellow :3 round green :3 wrinkled yellow :1 wrinkled green

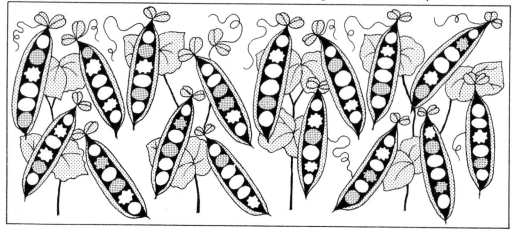

Figure 14.1 Dihybrid cross

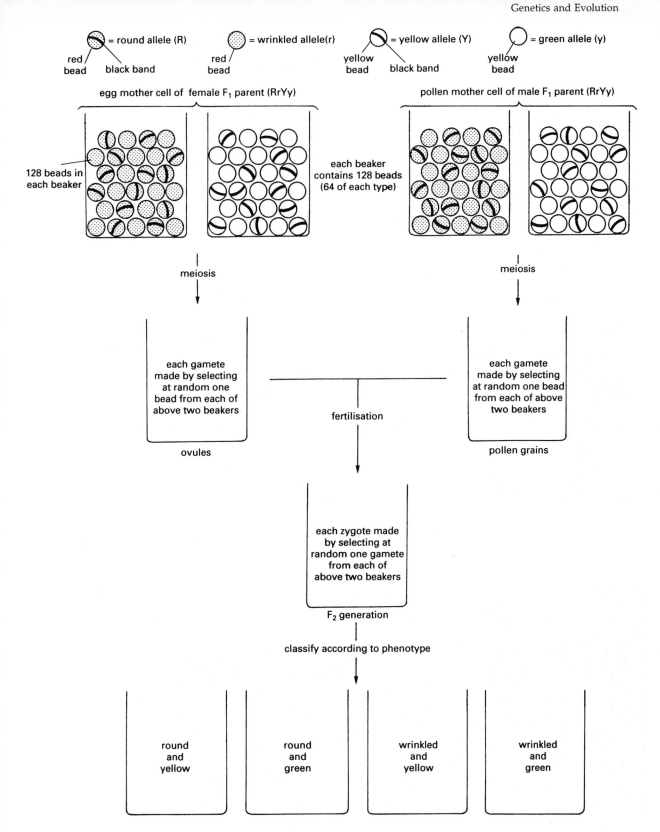

Figure 14.3   Bead model experiment of gamete formation

Recombination

In a dihybrid cross, two of the F_2 phenotypes resemble the original parents and two display new combinations of the characteristics. This process by which new combinations of parental characteristics arise is called **recombination** and the individuals possessing them are called **recombinants** (see figure 14.4).

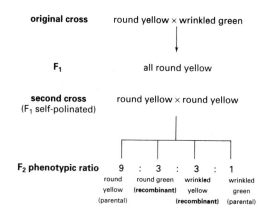

Figure 14.4   Recombinants

Use of fruit flies in breeding experiments

Table 14.1 lists some of the features that make the fruit fly (*Drosophila melanogaster*) suitable for use in genetics experiments.

feature	reason why feature is useful
short life cycle (about 10 days at 25°C)	allows the geneticist to study the transmission of an inherited characteristic through several generations in a short space of time
production of many offspring (female usually lays about 100 eggs)	allows the geneticist to make valid statistical analysis of results; (if only a few offspring were produced they might be unusual)
adult male differs from female in several obvious ways (see figure 14.5)	allows geneticist to easily identify and separate males from females before flies reach sexual maturity thus ensuring that females remain virgins until needed for next experiment

Table 14.1   Reasons for using fruit flies in genetics experiments

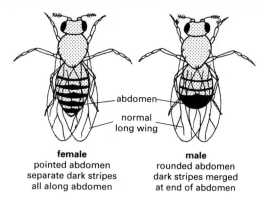

female
pointed abdomen
separate dark stripes
all along abdomen

male
rounded abdomen
dark stripes merged
at end of abdomen

Figure 14.5   Comparison of male and female fruit flies

Wild type

Amongst the members of a species, the type that occurs most commonly in wild populations is called the **wild type**. In *Drosophila*, the wild type has a grey body, long wings and red eyes. Each of these alleles is dominant to any other allele of the same gene. A wild type allele is often represented by the symbol '+'. However when alleles of two genes are being studied simultaneously, confusion can arise.

To avoid this, the following dihybrid crosses adopt the policy of representing each recessive allele by its lower case first letter and the wild type allele by the upper case of the same letter. Thus yellow body = y and grey body = Y; vestigial (tiny) wing = v and long wing = V.

Dihybrid cross in *Drosophila*

Precautions

1 The females must be virgins to ensure that the eggs have not been fertilised in advance by flies of unknown genotype.
2 Several of each sex must be used to allow for infertility or non-recovery from the anaesthetic.
3 After egg-laying, parents must be removed before the emergence of the next generation to avoid confusion.

In the following cross, true-breeding wild type males are crossed with females possessing yellow bodies and vestigial wings. The females are said to be **double recessive**. (The use of wild type females and double recessive males would be a genetically identical cross to this one.)

All of the F_1 generation produced are found to be wild type in appearance. Some F_1 males and females are then used as the parents of the second cross. Table 14.2 shows a typical set of results from this experiment.

	grey body long wing	yellow body long wing	grey body vestigial wing	yellow body vestigial wing
original cross	6♂			3♀
F₁	198			
parents of F₂	6♂ and 3♀ from F₁			
F₂	92	28	31	11

Table 14.2   Results of dihybrid cross

When the F₂ results are analysed statistically, they are found to represent a 9:3:3:1 ratio. It is concluded that the alleles for body colour and wing type are on different chromosomes which show independent assortment during meiosis.

Using the symbols stated above, the cross can be represented as follows:

original cross   YYVV × yyvv

gametes   all YV | all yv

F₁   all YyVv

second cross   YyVv × YyVv

gametes   YV, Yv, yV, yv | YV, Yv, yV, yv

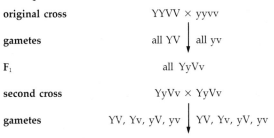

	♂ gametes			
	YV	Yv	yV	yv
♀ gametes YV	YYVV	YYVv	YyVV	YyVv
Yv	YYVv	YYvv	YyVv	Yyvv
yV	YyVV	YyVv	yyVV	yyVv
yv	YyVv	Yyvv	yyVv	yyvv

F₂ phenotypic ratio   9 long wing, grey body:
3 vestigial wing, grey body:
3 long wing, yellow body:
1 vestigial wing, yellow body

Dihybrid backcross (testcross)

If the F₁ flies produced in the previous cross are **backcrossed** (**testcrossed**) with double recessive flies (instead of being crossed with other F₁ flies), the following occurs:

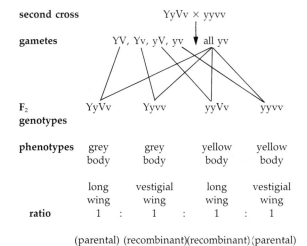

second cross   YyVv × yyvv

gametes   YV, Yv, yV, yv   all yv

F₂ genotypes   YyVv   Yyvv   yyVv   yyvv

phenotypes   grey body   grey body   yellow body   yellow body

long wing   vestigial wing   long wing   vestigial wing

ratio   1 : 1 : 1 : 1

(parental) (recombinant) (recombinant) (parental)

This F₂ generation consisting of 50% parental phenotypes and 50% recombinants occurs because the two genes involved are on separate chromosomes showing independent assortment.

Linkage

If each chromosome consisted of just one gene then Mendel's principle of independent assortment would hold true for all dihybrid crosses. However each chromosome is composed of a large number of genes and these

	long wing red eye	vestigial wing red eye	long wing brown eye	vestigial wing brown eye
original cross	6♂			3♀
F₁	181			
second cross (test cross)	6♂ from F₁			3♀
F₂	64 (parental)	36 (recombinant)	39 (recombinant)	61 (parental)

Table 14.3   Results of dihybrid test cross involving linked genes

do not behave independently of one another.

When a cross involves two alleles of two different genes located on the **same** chromosome, the two genes are **transmitted together** at meiosis and are said to be **linked**.

Consider the following example for *Drosophila* involving two linked genes. They are the gene for wing type (alleles long = V, vestigial = v) and the gene for eye colour (alleles red = B, brown = b). If the two genes were completely linked then a backcross (testcross) would fail to produce any F_2 recombinants. This is shown in figure 14.6.

However when this cross is carried out, the F_2 is found to contain some recombinants as shown in table 14.3 which gives a typical set of results. So how can this be explained?

Mechanism by which linked genes are separated

During meiosis when homologous chromosomes form pairs, **crossing over** may occur between adjacent chromatids at points called **chiasmata** (see page 62). If crossing over occurs between two genes, this separates alleles that were previously linked and allows them to recombine in new combinations. Figure 14.7 explains how the recombinants in table 14.3 arose.

Frequency of recombination

Since chiasmata can occur at any point along a chromosome, more crossing over (and therefore more recombination) occurs between two distantly located genes than between two that are close together.

The distance between two genes is therefore indicated by the percentage number of F_2 recombinants that result from a testcross.

This percentage is called the **recombination frequency** or **cross-over value** (**COV**) and is calculated as follows:

$$COV = \frac{\text{number of } F_2 \text{ recombinants}}{\text{total number of } F_2 \text{ offspring}} \times 100$$

Applying this formula to the results in table 14.3, the COV for the two genes = 75/200 × 100 = 37.5.

Chromosome maps

For convenience, a COV of 1% is taken to represent a distance of one unit on a chromosome. Thus the genes for long or vestigial wing and red or brown eye are located 37.5 units apart on one of *Drosophila's* chromosomes.

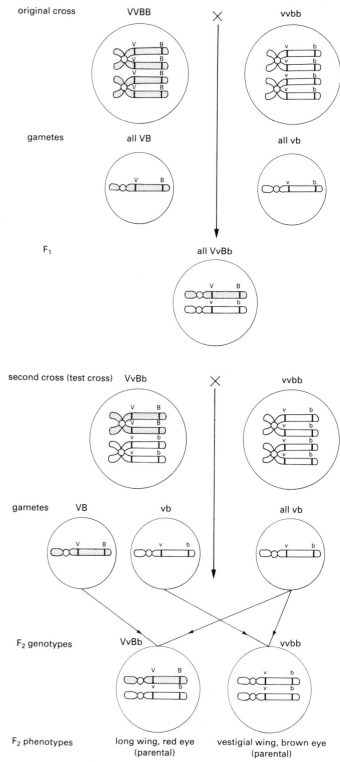

Figure 14.6 Complete linkage

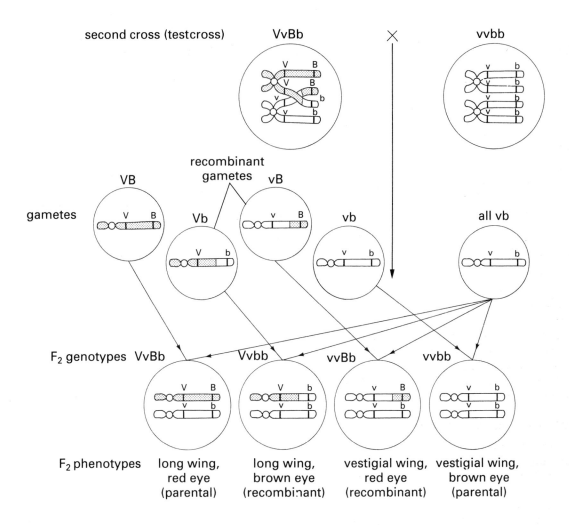

Figure 14.7   Separation of linked genes

The greater the COV, the longer the distance between two genes. The smaller the COV, the shorter the distance between two genes.

By carrying out several testcrosses, each involving two different linked genes at a time, it is possible to map the order and location of the genes on a chromosome. Figure 14.8 shows such a map for the imaginary genes given in table 14.4.

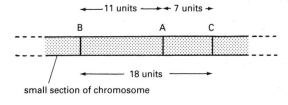

Figure 14.8   Chromosome map

linked gene pair	COV
AB	11
AC	7
BC	18

Table 14.4   Cross-over values

QUESTIONS

1 In tomato plants, round fruit shape is dominant to pear shape, and red fruit colour is dominant to yellow.

a) In diagrammatic form, follow to the F_2 generation a cross between a parent bearing pear shaped yellow fruit and one which is true-breeding for round red fruit. (Assume that the F_1 is self-pollinated and that the two genes are located on different chromosomes.)

b) In your Punnett square, underline FOUR individuals possessing different genotypes and phenotypes.

2 The four fruit flies in figure 14.9 are all offspring from the same parents.

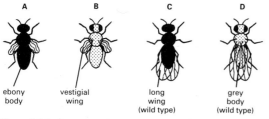

A	B	C	D
ebony body	vestigial wing	long wing (wild type)	grey body (wild type)

Figure 14.9

a) Fly **A** is homozygous recessive for the alleles of the genes affecting wing type and body colour. Give the fly a genotype.

b) Assume that in the cross, one parent had the same genotype as **A**. What gametes must this parent have made?

c) What types of gamete must have been produced by the second parent?

d) State the genotypes of the second parent and of each of flies **B**, **C** and **D**.

3 The following diagram shows a homologous pair of chromosomes bearing three marker genes R, S and T.

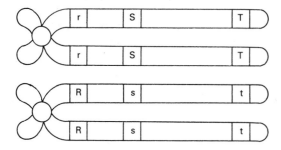

Figure 14.10

Rewrite the following sentences and complete the blanks.

a) Genes R, S and T, which have their loci on the same chromosome, are said to be _____ genes.

b) Crossing over, the process which can separate the alleles of such genes, occurs at points called _____.

c) There is a greater chance of crossing over occurring between genes _____ and _____ than between genes _____ and _____.

d) If a cross-over occurred between loci S and T then the recombinant gametes formed would be _____ and _____, and the parental gametes would be _____ and _____.

4 A cross was carried out between two maize plants, one true-breeding for brown pericarp and shrunken endosperm, the other true-breeding for white pericarp and full endosperm. F_1 plants (all white pericarp and full endosperm) were backcrossed against double recessive plants. The F_2 that resulted was:

    81 white pericarp, full endosperm:
    89 brown pericarp, shrunken endosperm:
    14 white pericarp, shrunken endosperm:
    16 brown pericarp, full endosperm.

a) Using symbols of your choice, present the information in diagrammatic form.

b) State which members of the F_2 are the recombinants.

c) (i) Calculate the percentage number of recombinants present in the F_2.
(ii) Why is this not 50%?

d) How many units apart on their chromosome are the genes that determine pericarp colour and endosperm type?

EXPERIMENTAL DESIGN QUESTION

Observations   The wild type form of *Drosophila*, the fruit fly, possesses a grey body and straight wings. Scientists have produced an unusual form of this animal which has a yellow body and curved wings.

Hypothesis   The genes for grey/yellow body colour and straight/curved wing type in *Drosophila* are located on different chromosomes and will show random assortment during gamete formation.

Problem   Outline an experimental procedure that would enable you to test the validity of this hypothesis.

DATA INTERPRETATION

In an experiment, three long-winged, red-eyed fruit flies were crossed with three vestigial-winged, purple-eyed flies. Three flies from the F_1 generation were backcrossed with three vestigial-winged, purple-eyed flies. The results in table 14.5 were obtained.

	long wing red eye	vestigial wing purple eye	long wing purple eye	vestigial wing red eye
cross 1 parents	3 (true-breeding)	3		
F₁ progeny	193			
cross 2 parents	3 (from F₁)	3		
F₂ progeny	88	87	13	12

Table 14.5

From the result it is concluded that, in *Drosophila*, vestigial wing (v) is recessive to long wing type (V) and purple eye (p) is recessive to red eye (P). The genes for wing type and eye colour must be linked since only 12.5% of the F_2 are recombinants.

Suitable backcrosses involving the gene for body colour, where black (b) is recessive to grey (B), reveal that this third gene belongs to the same linkage group as the first two genes. The recombination frequency between the genes for eye colour and body colour is found to be 6.0%. Since such values indicate the distances (in arbitrary units) between genes, the chromosome map must be either map 1 or map 2 as shown in figure 14.11

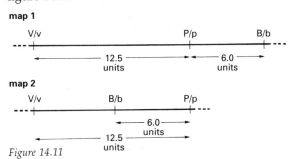

Figure 14.11

Map 1 is shown to be the correct one when a backcross involving the genes for wing type and body colour is found to produce 17.5% recombinants. The fact that this recombination frequency value is slightly lower than expected is due to the occurrence of an occasional double cross-over as shown in the following diagram.

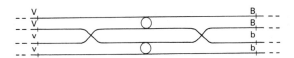

Figure 14.12

a) Upon what evidence is the conclusion that vestigial wing and purple eye are recessive alleles, based? (1)
b) Name the recombinant types given in the table of results (table 14.5). (1)
c) Since 12.5% of the F_2 generation in the table are recombinants, what process must have occurred between homologous chromosomes during meiosis? (1)
d) A backcross involving the linked genes for wing type and body colour is carried out.
(i) What should be the recombination frequency value?
(ii) What was it found to be by one direct experiment?
(iii) How does a double cross-over bring about such a discrepancy? (3)
e) Another member of this linkage group is the gene for bristle type. Reduced bristle (r) is recessive to normal bristle (R). The recombination value between the gene for eye colour and bristle type is 2.5% and between body colour and bristle type is 3.5%.
(i) Redraw the map to include this fourth gene.
(ii) Why is it necessary to know both of the cross-over values in order to plot gene R/r on the map? (2)
f) When another gene, W/w, is coupled with the gene for eye colour in the backcross WwPp × wwpp, no recombinants are formed at all.
(i) Where must gene W/w be situated?
(ii) Give the genotypes of the members of the backcross whose recombinant offspring would help to verify your prediction. (2)

(10)

15 Sex linkage

Sex determination

In the nucleus of every normal human **somatic** (body) cell, there are 46 chromosomes. These exist as 22 homologous pairs of **autosomes** and one pair of **sex chromosomes**. This latter pair determine an individual's sex.

In the female, the sex chromosomes make up a homologous pair, the **X** chromosomes. Thus a human female has the chromosome complement: **44 + XX**.

In the male, the sex chromosomes make up an unmatched pair, an X chromosome and a smaller **Y** chromosome. Thus a human male has the chromosome complement: **44 + XY**.

An egg mother cell is said to be **homogametic** since every egg formed by meiosis contains an X chromosome. A sperm mother cell is **heterogametic** since half of the sperm formed contain an X chromosome and half a Y chromosome. When the nucleus of a sperm fuses with the nucleus of an egg at fertilisation, the sex of the zygote is determined by the type of sex chromosome carried by the *sperm*.

In the locust, the female has an even number of chromosomes, 18 (16 + XX) and the male has an odd number, 17 (16 + X). Thus half the sperm do receive an X chromosome, the other half do not, and sex is determined as before.

The female is not always the homogametic sex. For example in birds and butterflies, the female is heterogametic (XY) and the male is homogametic (XX).

Sex linkage

Although sex chromosomes behave as a homologous pair at meiosis, an X and a Y chromosome do not make up a truly homologous pair. In humans for example, the larger X chromosome carries many genes that are not

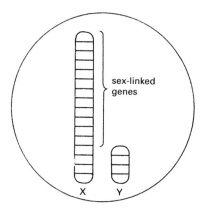

Figure 15.1   Sex-linked genes

present in the smaller Y chromosome. This is shown in figure 15.1.

These genes are said to be **sex-linked**. When a sex-linked gene occurs on the X chromosome but not on the Y, and that X chromosome meets a Y at fertilisation, then the sex-linked characteristic (whether dominant or recessive) will be expressed in the phenotype of the organism produced. This is because the Y chromosome has no allele at the equivalent gene locus to offer dominance.

Eye colour in *Drosophila*

Drosophila possesses the same mechanism of sex determination as humans. Although the Y chromosome is not smaller than the X, it is a different shape and carries very few genes. The X chromosome therefore carries many sex-linked genes. One of these determines eye colour.

The allele for red eye colour (R) is dominant to the allele for white eye (r). Figure 15.2 shows a cross between a white-eyed female and a red-eyed male followed through to the F_2 generation.

A simpler version of this cross is shown in

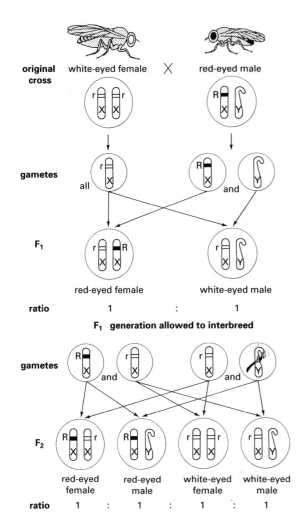

Figure 15.2   Sex linkage in Drosophila

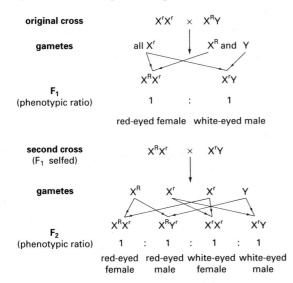

Figure 15.3   Sex-linked inheritance in Drosophila *using symbols*

figure 15.3 where the sex chromosomes are represented by X and Y and the alleles of the sex-linked gene by the superscripts of R and r.

Since this cross involves a sex-linked gene, the F_2 generation does not show the phenotypic ratio of 3:1 typical of a normal monohybrid cross.

Red-green colour-blindness

In humans, normal colour vision (C) is dominant to red-green colour-blindness (c). These are the alleles of a sex-linked gene on the X chromosome. The five possible genotypes and their phenotypes are given in table 15.1.

genotype	phenotype
$X^C X^C$	female with normal colour vision
$X^C X^c$	female (carrier) with normal colour vision
$X^c X^c$	female with colour-blindness (very rare e.g. 0.5% of European population)
$X^C Y$	male with normal colour vision
$X^c Y$	male with colour-blindness (more common e.g. 8% of European population)

Table 15.1   Red-green colour-blindness in humans

Heterozygous females are called **carriers** because, although unaffected themselves, they pass the allele on to 50% of their offspring. On average therefore, 50% of a carrier female's sons are colour-blind. This is shown in figure 15.4.

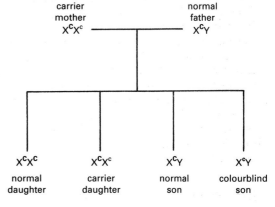

Figure 15.4   Colour-blindness cross using symbols

Red-green colour-blindness is rare in females since two recessive alleles must be inherited. It is more common in males where only one is needed.

Haemophilia

Haemophilia is a disease involving defective blood clotting. It is caused by a recessive allele carried on the X chromosome and is therefore sex-linked.

Y-linked genes

Hairy ears in humans is a trait which is transmitted from father to sons but never to daughters. It is possible therefore that it is determined by a gene which occurs on the Y but not on the X chromosome.

QUESTIONS

---

1 In diagrammatic form, follow through to the F_2 generation a cross between a white-eyed male *Drosophila* and a homozygous red-eyed female, where the F_1 are allowed to interbreed.
2 Using the same format as figure 15.4, show the possible results of a cross between a colourblind male and a carrier female.
3 Haemophilia occurs in the family tree shown in figure 15.5.

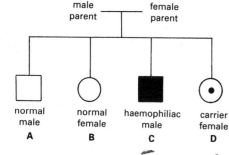

Figure 15.5

a) Using the convention X^H (normal allele), X^h (haemophilia allele) and Y (no allele), give the genotypes of the offspring, **A, B, C** and **D**.
b) Give the genotype and phenotype of each parent.
c) If **C** marries a normal female, what proportion of their sons are likely to be haemophiliacs?
4 In cats, coat colour is determined by a sex-linked gene where allele B is black and allele G is ginger. In heterozygotes, BG results in tortoiseshell as shown in table 15.2.

phenotype	female genotype	male genotype
black	$X^B X^B$	$X^B Y$
ginger	$X^G X^G$	$X^G Y$
tortoiseshell	$X^B X^G$	

Table 15.2

a) Why is no tortoiseshell male given in the table?
b) Show in diagrammatic form, TWO crosses that would result in all the female offspring being tortoiseshell.
5 In poultry, the gene for plumage colour is sex-linked. The allele for white feathers (W) is dominant to the allele for red feathers (w). In an attempt to demonstrate that this gene is sex-linked, the following cross was set up.

$$X^w X^w \qquad \times \qquad X^W Y$$
white male   red female

a) Explain why the outcome of this cross would NOT show that the gene is sex-linked.
b) Present in diagrammatic form a cross that would verify that the gene is sex-linked. Explain your answer.

16 Mutation

A **mutation** is a sudden change in the structure or amount of an organism's genetic material. It varies in form from a tiny change in the DNA structure of a gene to a large scale alteration in chromosome structure or number. When such a change in genotype produces a change in phenotype, the individual affected is called a **mutant**.

Change in chromosome number

Non-disjunction

Sometimes during meiosis a spindle fibre fails and the members of a homologous pair of chromosomes do not become separated. This is called **non-disjunction** (see figure 16.1). As a

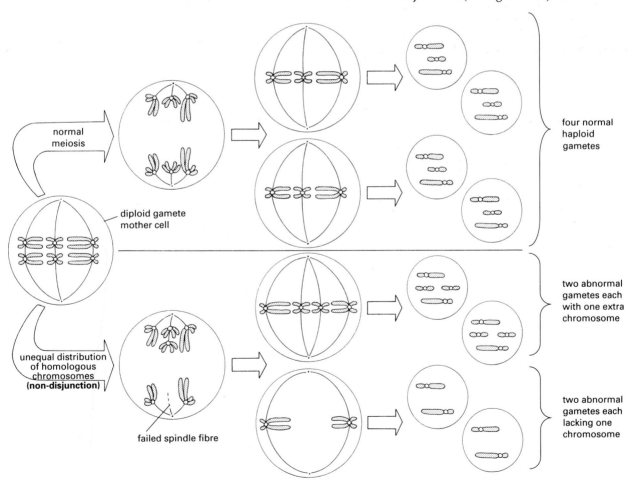

Figure 16.1   Effect of non-disjunction

result two of the gametes formed receive an extra chromosome and two gametes lack that chromosome.

If non-disjunction of chromosome 21 occurs in a human egg mother cell, then an abnormal egg (n=24) may be fertilised by a normal sperm (n=23) forming a zygote (2n=47). The extra chromosome can be seen in the **karyotype** (display of matched chromosomes) shown in figure 16.2. The mutant individual suffers **Down's Syndrome** (**mongolism**) which is characterised by mental retardation and distinctive physical features.

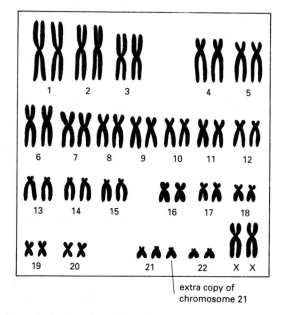

Figure 16.2 *Down's syndrome karyotype*

Polyploidy

The single haploid set of chromosomes typical of a species is called its **genome**. This is often represented by a symbol (e.g. A). **Polyploidy** is the increase in number of the species' genome by three or more times.

Polyploidy from a single species
Some polyploids (e.g. AAA, AAAA, etc.) contain more than two genomes all derived from a single species (AA). Such polyploids arise in a number of ways.

All of the chromosomes in a gamete mother cell of AA may undergo non-disjunction during meiosis resulting in the formation of two diploid gametes (see figure 16.3). Subsequent fusion of a diploid gamete (AA) with a normal haploid one (A) results in a **triploid** organism (AAA) being formed. Although capable of mitosis and normal

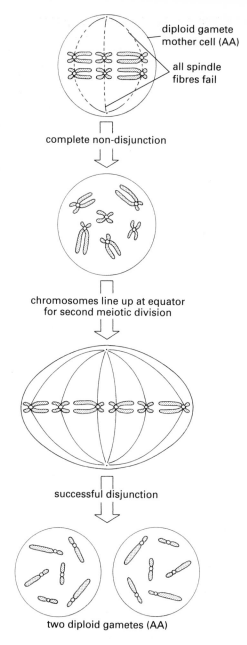

Figure 16.3 *Effect of complete non-disjunction*

growth, this is sterile since its chromosomes cannot form homologous pairs during meiosis. Many varieties of daffodils and other bulb plants are sterile triploids which survive by asexual reproduction.

Fusion of two **diploid** gametes produces a **tetraploid** (AAAA) which may be fertile. AAAA can also arise by spindle failure during mitosis in a cell of AA. Once formed, the polyploid cell divides normally to form a polyploid side shoot on the diploid plant.

Polyploids from more than one species

Some polyploids (e.g. AABB) contain multiple sets of chromosomes derived from more than one species. In this case a gamete containing genome A from species AA fuses with a gamete containing genome B from species BB to produce a **sterile hybrid** AB. AB survives by asexual reproduction until eventually a mitotic spindle failure in one of its cells results in **tetraploid** AABB being formed. AABB is fertile since its chromosomes can form homologous pairs at meiosis.

Thus polyploidy is a method by which a new species can be produced 'overnight'. It is thought that rice grass (*Spartina townsendii*) arose in this way as follows:

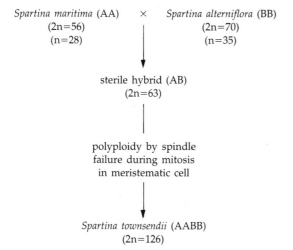

Economic significance of polyploidy

Polyploidy is very rare in animals but common in plants. Polyploid plants usually show an increase in **size, vigour** and **resistance to disease.** *Spartina townsendii*, for example, is hardier and more fertile than either of its ancestors and it used to begin cultivation of poor salty land reclaimed from the sea.

Many crop plants such as bananas, apples, tomatoes, sugar cane, coffee and wheat are polyploid and give a bigger yield than their diploid relatives.

Polyploidy can be induced by the chemical **colchicine** which prevents spindle formation during mitosis. When colchicine is removed the cells (now polyploid) divide normally. Scientists use this method to produce new varieties of crop plant.

Change in structure of one chromosome

This type of mutation involves a change in the **number** or **sequence** of genes in a chromosome. Such a change is most likely to occur when chromatids break and rejoin during crossing over at meiosis. Figure 16.4 shows four ways in which this can happen.

Tunicate locus in corn

Podcorn (see figure 16.5) is thought to be very similar to the ancestral form of maize (Indian corn). Each of its kernels (grains) is surrounded by a husk. This characteristic represents the phenotypic expression of the dominant allele Tu (tunicate means wrapped in a tunic) of a locus situated on chromosome 4. Thus true-breeding podcorn has the genotype TuTu.

Popcorn is a closely related form of maize used in the confectionery industry. Each of its kernels lacks a husk. This represents the phenotypic expression of the recessive allele (tu) at the same locus. Thus true-breeding popcorn has the genotype tutu.

The tunicate locus is now known to be a compound locus ('supergene') consisting of three separate components (genes) thought to have arisen by **duplication**. Since each of the dominant components of the supergene, Tu, contributes in part towards husk formation, it is possible to produce hybrids between pod and popcorn which have partly formed husks (i.e. semi-tunicate).

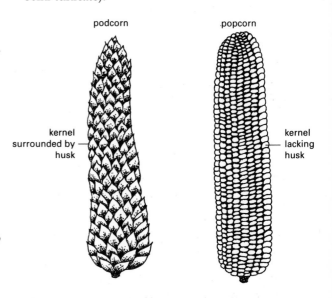

Figure 16.5 Podcorn and popcorn

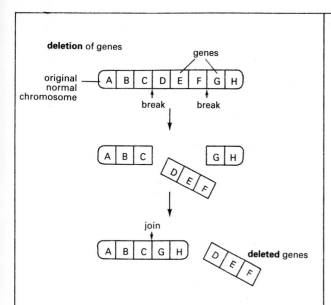

deletion of genes

significance: Genes unattached to centromere are lost. Condition usually **lethal** (deadly) since mutant is unable to code for certain essential proteins.

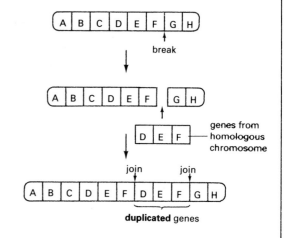

duplication of genes

significance: One of duplicated genes may be able to mutate and evolve a new beneficial function.

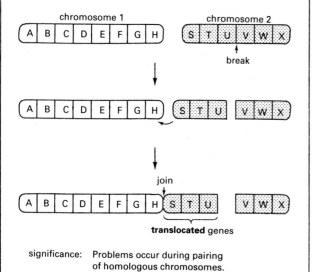

translocation of genes from one chromosome to another non-homologous one

significance: Problems occur during pairing of homologous chromosomes. Gametes formed are often not viable.

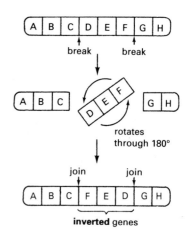

inversion of genes

significance:- Cross-overs involving inverted genes lead to formation of non-viable gametes. However inverted genes can be passed on if they remain as a group unaltered by crossing over. If they happen to be a group of advantageous alleles, then this so-called 'supergene' is transmitted intact from generation to generation.

Figure 16.4   Changes in structure of a chromosome

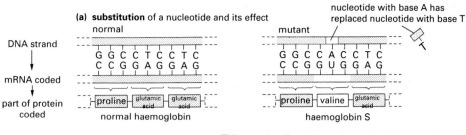

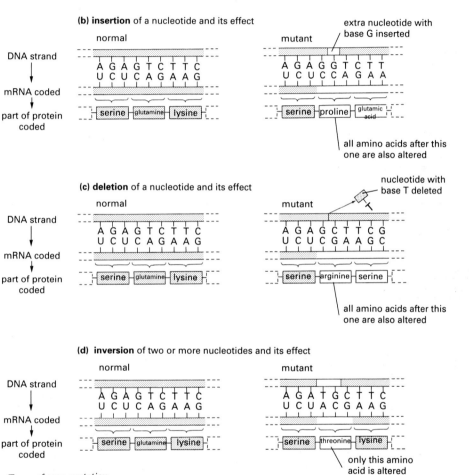

Figure 16.6 Types of gene mutation

Alteration of base type or sequence in DNA

This type of mutation involves a change in one or more of the **nucleotides** in the DNA strand. In each of the examples shown in figure 16.6, the codon for a particular amino acid has become altered leading to a change in the protein that is synthesised.

For a protein to function properly it must possess the correct sequence of amino acids. A minor change (i.e. one different amino acid) caused by a substitution or an inversion may be tolerated. However the complete 'frameshift' and subsequent misreading of a large portion of the gene's DNA caused by an insertion or a deletion normally leads to a non-functional protein.

If an enzyme is made which is no longer able

organism	characteristic controlled by normal gene	mutant characteristic resulting from gene mutation
fruit fly	long wing grey body red eye	vestigial wing yellow body white eye
human	normal blood clotting secretion of normal mucus in lung normal haemoglobin and biconcave red blood cells	haemophilia secretion of viscous mucus which blocks bronchioles (cystic fibrosis) haemoglobin S and sickle-shaped red blood cells
mouse	brown coat	white coat (albino)
maize	green leaves containing chlorophyll	albino leaves lacking chlorophyll

Table 16.1   Mutant characteristics

organism	mutant characteristic	mutation rate (mutations at gene locus/million gametes)	chance of new mutation occurring
pneumonia bacterium	resistance to penicillin	0.1	1 in 10 000 000
fruit fly	ebony body white eye	20 40	1 in 50 000 1 in 25 000
mouse	albino coat	10	1 in 100 000
human	haemophilia muscular dystrophy	45 80	1 in 22 000 1 in 12 500

Table 16.2   Mutation frequency

to catalyse some essential step in a biochemical pathway, an intermediate metabolite may accumulate with disastrous results (see phenylketonuria, page 141).

Since most proteins are indispensable to the organism, most gene mutations produce an inferior version of the phenotype (see table 16.1). If this results in death (an albino plant, for example, cannot photosynthesise) then the altered gene is said to be **lethal**.

Frequency of mutation

In the absence of outside influences, gene mutations arise *spontaneously* and at *random* but only occur *rarely*. The mutation rate of a gene is expressed as the number of mutations that occur at that gene locus per million gametes. Mutation rate varies from gene to gene, and from species to species, as shown in table 16.2.

The vast majority of mutant alleles are recessive. Therefore a newly formed mutant allele fails to be expressed phenotypically until a reshuffle of alleles eventually pairs it up with a similar mutant allele in some future generation. However a few mutant alleles do show up immediately because they are either dominant (e.g. achondroplasia, see question 8 on page 88) or sex-linked (e.g. haemophilia).

Mutagenic agents

Mutation rate can be artificially increased by **mutagenic agents**. These include certain chemicals (e.g. mustard gas) and various types of radiation (e.g. gamma rays, X-rays and UV light). The resultant mutations are said to be **induced**.

The effect of radiation on seeds is shown in figure 16.7 where an increase in level of radiation is seen to bring about an increase in rate of gene mutation.

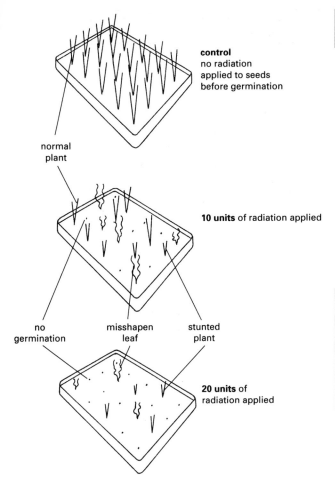

normal plant

control
no radiation
applied to seeds
before germination

10 units of radiation applied

no
germination

misshapen
leaf

stunted
plant

20 units of
radiation applied

Figure 16.7   Effect of radiation on seeds

Mutation as a source of variation

Mutation is the only source of *new* variation. It is the process by which new alleles of genes are produced. Without mutation, all organisms would be homozygous for all genes and no variation would exist.

Most mutations are harmful or even lethal. However, on very rare occasions, there occurs by mutation a mutant allele which confers some *advantage* on the organism that receives it. Such mutant alleles (which are better than the originals) provide the alternative choices upon which natural selection can act. They are therefore considered to be the raw material of evolution (see page 90).

(see page 90)

QUESTIONS

1 With the aid of an example distinguish clearly between the terms *mutation* and *mutant*.
2 In the following two telegrams, a small error alters the sense of the message.
   a) *Intended*: John walked to the bus.
   Actual: John talked to the bus.
   b) *Intended*: The two strings were untied.
   Actual: The two strings were united.
   To which type of mutation shown in figure 16.6 is each of these analogous?
3 **a)** Each of the homologous pairs of chromosomes shown in figure 16.8 was seen under a microscope during meiosis in two cells. Which type of chromosome mutation had occurred in each case?

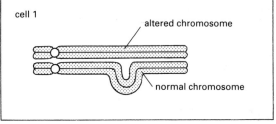

cell 1

altered chromosome

normal chromosome

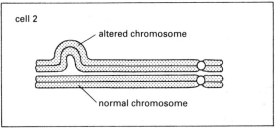

cell 2

altered chromosome

normal chromosome

Figure 16.8

b) Which of these would be more likely to prove lethal to the organism that inherited the affected chromosome?
4 The following diagram shows the genetic relationships between some species of *Primula*.

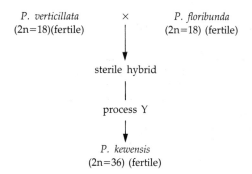

P. verticillata   ×   P. floribunda
(2n=18)(fertile)   (2n=18) (fertile)

sterile hybrid

process Y

P. kewensis
(2n=36) (fertile)

a) What is the diploid (2n) chromosome complement for the sterile hybrid?
b) Explain why this hybrid is sterile.
c) Name process Y.

5 Table 16.3 refers to three species of the *Brassica* group of plants.

scientific name	common name	diploid chromosome number (2n)	haploid chromosome number (n)
Brassica oleracea	cabbage	18	
Brassica rapa	turnip		10
Brassica rapobrassica	swede	38	

Table 16.3

a) Copy and complete table 16.3.
b) Scientists consider cabbage and turnip to be the original parents of swede but a cross between cabbage and turnip produces a sterile hybrid.
(i) State the diploid chromosome number of this hybrid.
(ii) Explain why it is sterile.
(iii) Construct a hypothesis to explain how swede could have arisen from this hybrid.
(iv) Given a supply of colchicine and several specimens of the sterile hybrid, describe briefly how you would test your hypothesis experimentally.

6 Explain how the information in figure 16.9 supports the following statement. '*After 30 years in a woman's ovary, gamete mother cells seem to be more liable to suffer non-disjunction during meiosis.*'

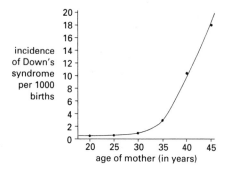

Figure 16.9

7 Although gene mutations are normally rare, the gene controlling colour of grains in maize mutates as often as once in 2000 gametes on average. Express this as a mutation rate. (See table 16.2 for help.)

8 Achondroplasia (a form of dwarfism in humans) is controlled by a dominant mutant allele (A). All achondroplastic dwarfs are heterozygous (Aa) since the homozygous condition is lethal. Imagine that each of the following sets of parents produce a dwarf child.
(i) Aa × Aa   (ii) Aa × aa   (iii) aa × aa
a) In which family must a gene mutation definitely have occurred?
b) From studies, it is now known that mutation rate of this gene is 14 per million gametes. Look back to table 16.2 and then express this figure as a 1 in _____ chance of a new mutation occurring.

9 In addition to affecting a living organism's sex cells, a mutagenic agent can alter normal body (somatic) cells.
a) Name TWO mutagenic agents.
b) Suggest why a mutation affecting the germ line (i.e. sex cells) is of much greater significance than a somatic mutation.

10 Following a long and detailed study, scientists reported in early 1990 that a definite link exists between exposure to radiation of men working at one of Britain's nuclear processing plants and the incidence of leukaemia amongst their children.
Construct a hypothesis to account for these findings.

DATA INTERPRETATION

An autopolyploid contains three or more genomes that have originated from one species. An allopolyploid possesses three or more genomes from more than one species.

For many years it was thought that *Triticum monococcum* (see table 16.4) had, by multiplication of its genome, given rise directly to modern bread wheat *Triticum vulgare*. However when a cross between species 1 and 3 was carried out, the resultant hybrid was found

species	wheat plant	chromosome number		genomes	
		n	2n	early theory	current theory
1	Triticum monococcum	7	14	AA	AA
2	Triticum durum	14	28	AAAA	AABB
3	Triticum vulgare	21	42	AAAAAA	AABBCC

Table 16.4

to be sterile. When its karyotype was inspected, genome analysis showed it to be AABC and not AAAA as expected. The simplified family tree in figure 16.10 shows how modern bread wheat is thought to have arisen.

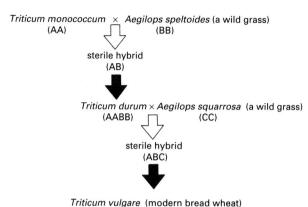

Triticum monococcum × *Aegilops speltoides* (a wild grass)
(AA)   (BB)

sterile hybrid
(AB)

Triticum durum × *Aegilops squarrosa* (a wild grass)
(AABB)   (CC)

sterile hybrid
(ABC)

Triticum vulgare (modern bread wheat)
(AABBCC)

Figure 16.10

a) The hybrid AB is sterile yet is close relative AABB is fertile. Explain why. (2)

b) What experimental evidence verifies that *Triticum vulgare* and *Triticum monococcum* are correctly classified as two distinct species? (1)

c) What process in the family tree diagram is represented by (i) an unshaded arrow, (ii) a shaded arrow? Has *Triticum vulgare* arisen by allopolyploidy or autopolyploidy? Briefly explain your answer. (4)

d) How many chromosomes are present in the genome of *Aegilops speltoides*? (1)

e) When *Triticum vulgare* is crossed with a certain species of rye, a sterile hybrid is formed which, following polyploidy, becomes a fertile plant called *Tricale* ($2n = 56$). By means of a keyed diagram (family tree) work out the chromosome complements of the sterile hybrid and the parental rye plant. (2)

(10)

What you should know (CHAPTERS 12–16)

1 **Sexual reproduction** is the means by which **genetic variation** is maintained in a population.

2 **Meiosis** is the process by which **haploid gametes** are formed.

3 During meiosis, **new combinations** of existing alleles arise by **independent assortment** of chromosome and **crossing over** between homologous chromosomes.

4 If an F_2 generation with a phenotypic ratio of **9:3:3:1** is obtained as a result of a dihybrid cross where the F_1 is **selfed**, the two genes involved must be located on **different** chromosomes.
However when the expected ratio of 9:3:3:1 is **not** obtained, this indicates that the two genes are located on the **same** chromosome (i.e. are **linked**).

5 If an F_2 generation with a ratio of **1:1:1:1** is obtained as a result of a dihybrid cross when the F_1 is **back crossed** (test crossed) to the double recessive, the two genes must be located on different chromosomes.

However when the expected ratio of 1:1:1:1 is **not** obtained, this also indicates that the two genes are located on the **same** chromosome (i.e. are **linked**).

6 Linked genes become **separated** if crossing over occurs between them. This produces **recombinant** gametes.

7 Since the **distance** between two linked genes is directly related to the **frequency of recombination** between them, recombination values can be used to construct gene maps of chromosomes.

8 Genes present on an X but not on a Y chromosome are **sex-linked**.

9 **Mutations** are **alterations** in genotype which involve a change in structure, or number of chromosomes, or base type, or sequence of a gene's DNA.

10 Mutations occur **rarely** and at **random**. Their frequency can be increased artificially by **mutagenic agents**.

11 Mutations are the only **source of new variation** and provide the raw material for evolution.

17 Natural selection

Historical background

In 1858 Charles Darwin and Alfred Wallace published a joint paper in which they suggested that the main factor producing evolutionary change is **natural selection**.

A year later Darwin amplified this view in his famous book *The Origin of Species* as follows:

a) Organisms tend to produce *more offspring* than the environment will support.

b) A struggle for existence follows and a large number of these offspring *die* before completing their life cycle. This is a result of overcrowding, competition and lack of food.

c) Members of the same species are not identical but show *variation* in all characteristics.

d) Those offspring whose phenotypes are *better suited* to their immediate environment will, in time, have a better chance of being among the *survivors*, reaching reproductive age and passing on the favourable characteristics to their offspring.

e) Those offspring whose phenotypes are *less well suited* to their immediate environment will *die* before reaching reproductive age.

f) Over a period of time, the best suited phenotypes will predominate in the population i.e. the *fittest* will *survive*.

Darwin called this 'weeding out' process **natural selection**. Since environmental conditions are constantly changing, natural selection is always favouring the emergence of new forms.

Darwin formulated his theory without any knowledge of genetics because Mendel's laws, although published in 1865, lay unnoticed until 1900. Darwin always felt that the weakest point of his argument lay in the lack of an adequate explanation of how variation arose and how it was transmitted from generation to generation. If Darwin had had the benefit of the information covered in chapters 12–16 of this book, the development of his theory would have been hastened.

Natural selection in action

Most mutations produce inferior versions of the original gene. However in each of the following examples of populations there occurs a **mutant allele** of a gene which allows adaptation to the changing environment. This allele gives the mutant form of the organism a **selective advantage**. The change in the environment may be brought about by an abiotic (non-living) factor such as atmospheric pollution or by a biotic (living) factor such as disease.

Industrial melanism in peppered moth (*Biston betularia*)

The existence of two or more forms within the same species is called **polymorphism**. One form of the peppered moth is light brown with dark speckles; the other is completely dark (**melanic**) in colour. They differ by only one allele of the gene controlling the formation of dark pigment (melanin). Both forms of the moth fly by night and rest on the bark of trees during the day.

Prior to the industrial revolution in the 1800s, the light form was common throughout Britain and the darker form which occasionally arose by mutation was very rare indeed.

Surveys in the 1950s (see figure 17.1) showed that the pale form was most abundant in non-industrial areas whereas the dark form was abundant in areas suffering from heavy industrial air pollution. Experiments and direct observations strongly support the following explanation of these findings.

In non-polluted areas, the tree trunks are covered with pale-coloured **lichens** and the light-

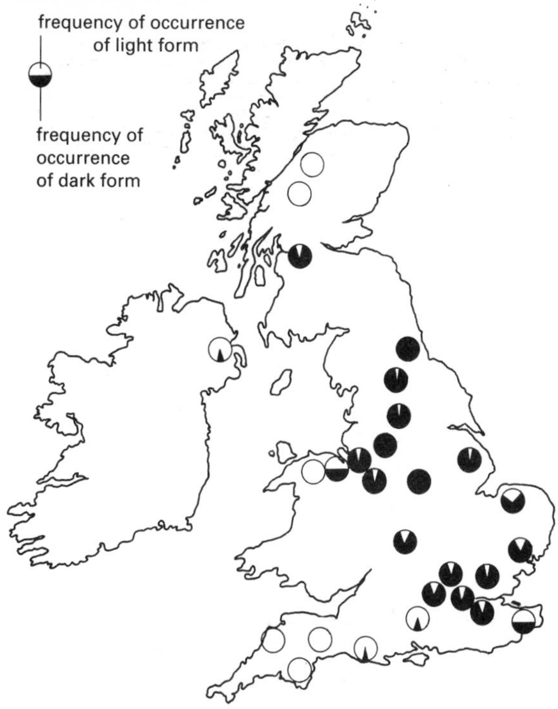

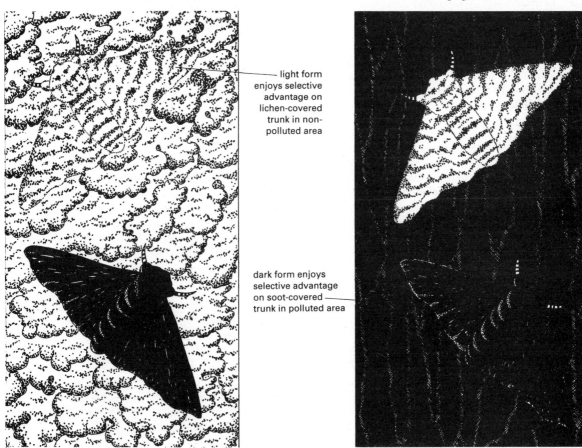

frequency of occurrence
of light form

frequency of
occurrence
of dark form

Figure 17.1   Frequencies of two forms of peppered moth

coloured moth is well **camouflaged** against this pale background (see figure 17.2). However the dark form is easily seen and eaten by predators such as thrushes.

In polluted areas, toxic gases kill the lichens and **soot** particles darken the tree trunks. As a result the light-coloured moth is easily seen whereas the dark one is well hidden and favoured by natural selection.

Nowadays, in polluted areas currently undergoing cleaning up campaigns in response to the Clean Air Acts, the pale form is being naturally selected at the expense of the melanic form which is losing its selective advantage. As the environment continues to change and more pale moths survive and breed, the situation which existed before the industrial revolution may return.

Sickle cell trait

Sickle cell anaemia is a genetically transmitted disease of the blood. It is caused by the presence of abnormal **haemoglobin S** which occurs as a result of a mutation (see page 85). In the

light form enjoys selective advantage on lichen-covered trunk in non-polluted area

dark form enjoys selective advantage on soot-covered trunk in polluted area

Figure 17.2   Natural selection in peppered moth

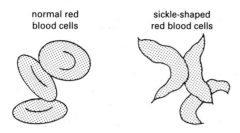

normal red blood cells   sickle-shaped red blood cells

Figure 17.3   Two types of red blood cell

	cross		
	HH × HH	HH × HS	HS × HS
possible result in non-malarial region	HH, HH, HH, HH (100% of offspring survive)	HH, HH, HS, HS (100% of offspring survive)	HH, HS, HS, ⊗ (75% of offspring survive)
possible result in malarial region affected by outbreak of disease	⊗, ⊗, ⊗, ⊗ (0% of offspring survive)	⊗, ⊗, HS, HS (50% of offspring survive)	⊗, HS, HS, ⊗ (50% of offspring survive)

Table 17.1   Genetics of sickle-cell anaemia

discussion that follows, H represents the allele for normal haemoglobin, and S represents the allele for haemoglobin S.

The blood of people who are homozygous (SS) for the mutant allele contains haemoglobin S (which is inefficient at carrying oxygen) and **sickle-shaped red blood cells** (see figure 17.3). These stick together and interfere with blood circulation. This causes severe anaemia, damage to vital organs and, in the majority of cases, death.

Allele H is **incompletely dominant** to allele S. In heterozygotes (HS), allele S is therefore partially expressed. This milder condition, known as the **sickle cell trait**, is characterised by about a third of the person's haemoglobin being type S. However the red blood cells are normal and the slight anaemia that results does not prevent modest activity.

Allele S is rare in most populations since it is semi-lethal. However, in some parts of Africa, up to 40% of the population are genotype HS. Comparison of the two maps in figure 17.4 shows that a correlation exists between incidence of **malaria** and a high frequency of allele S.

This is because sickle cell trait sufferers are resistant to malaria whereas people with normal haemoglobin are not. Thus, in malarial regions, natural selection favours people with genotype HS over those with HH who may die during

serious outbreaks of the disease (see table 17.1). However HS loses its selective advantage in non-malarial areas.

Resistance to antibiotics

Some mutant forms of bacteria are *resistant* to antibiotics. The experiment shown in figure 17.5 demonstrates the occurrence of natural selection. In this case an **induced mutant** is found to be resistant to the antibiotic.

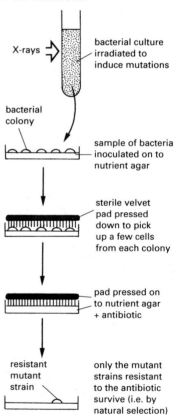

X-rays → bacterial culture irradiated to induce mutations

bacterial colony

sample of bacteria inoculated on to nutrient agar

sterile velvet pad pressed down to pick up a few cells from each colony

pad pressed on to nutrient agar + antibiotic

resistant mutant strain

only the mutant strains resistant to the antibiotic survive (i.e. by natural selection)

Figure 17.5   Isolation of resistant bacteria

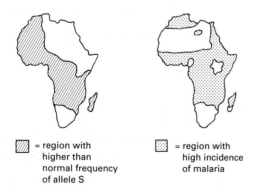

▨ = region with higher than normal frequency of allele S

▦ = region with high incidence of malaria

Figure 17.4   Correlation between allele S and malaria

Spontaneous mutations are also occurring all the time. As a result, a few bacteria within a population may already possess resistance alleles to an antibiotic even though they have never been exposed to it. This can lead to the situation illustrated in figure 17.6.

Already the usefulness of many antibiotics has been greatly reduced by the existence of resistant bacteria. Many disease-causing strains of *Staphylococcus* make the enzyme penicillinase which enables them to digest and resist penicillin. Small wonder that pharmaceutical companies are constantly trying to discover and develop new antibiotics.

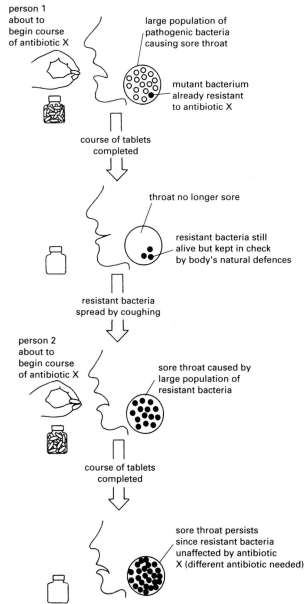

Figure 17.6   Spread of resistant bacteria

Resistance to insecticides

DDT is a poisonous chemical which has been widely used against many insects. These include mosquitoes which carry malaria and yellow fever. Within a few years of use, many mutant forms of insects resistant to the insecticide 'appeared'.

These had not arisen in response to the chemical. A tiny number of resistant mutants just happened to be present within the natural insect populations, or they arose later by chance.

When the spray was applied, the vast majority of non-resistant insects died and the resistant mutants suddenly enjoyed a selective advantage and multiplied. Under such circumstances natural selection enables the mutant to replace its wild type relatives.

Many new insecticides have been developed in recent years but strains of insects resistant to these chemicals are now emerging.

Heavy metal tolerance in grasses

Soil in the waste tips around mines where ores were once extracted contain high concentrations of heavy metals. These include copper and lead. Although such concentrations are normally toxic to plants and animals, some varieties of grass species are able to tolerate and colonise these polluted sites.

Look at figure 17.7 which refers to two varieties of the grass *Agrostis tenuis* found growing around a Welsh copper mine.

Table 17.2 shows results from a competition experiment where seeds from both types of this

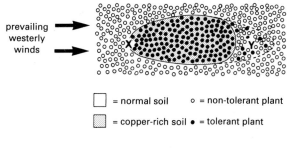

☐ = normal soil   o = non-tolerant plant

▨ = copper-rich soil   ● = tolerant plant

Figure 17.7   Heavy metal tolerance

	percentage survival	
	seeds from tolerant plants	seeds from non-tolerant plants
soil rich in copper	98	0
normal soil	23	89

Table 17.2   Competition experiment results

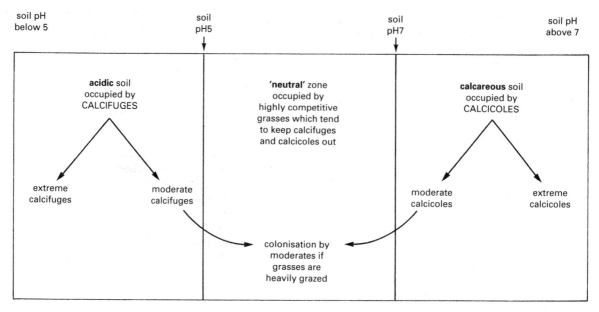

Figure 17.8 *Distribution of calcifuges and calcicoles*

grass were grown together in the two types of soil. From such studies it is now known that **copper tolerance** in *Agrostis tenuis* is an inherited characteristic (involving mutant alleles of several genes).

This is a further example of **natural selection** in action. Only those plants possessing the genetic combination that results in copper tolerance survive on toxic soil. They pass this characteristic on to their offspring which are in turn selected.

At point X (the upwind end of the mine), the boundary between the tolerant and non-tolerant populations is sharp because any pollen or seeds carrying non-tolerant genes blown over the boundary fail to produce tolerant plants.

In zone Y (beyond the downwind end of the mine), both varieties of plant occur. Although most of the seeds blown into this area are of the tolerant type, tolerant plants are found to be in the minority because they are unable to compete well with the non-tolerant variety when the soil's copper concentration is low.

Calcifuge and calcicole pairs

A group of plants which has a wide ecological range may evolve a pair of distinct ecotypes each adapted to a specific habitat. **Calcifuges**, for example, are plants which grow best on *acidic* soils whereas their **calcicole** relatives grow best on *calcareous* (calcium-rich) soils (see figure 17.8).

Since this difference in soil preference is genetically determined it is passed on to subsequent generations. Natural selection continues to favour calcifuges (e.g. *Viola palustris*, the marsh violet) on the acidic soils of marshes; calcicoles (e.g. *Viola hirta*, the hairy violet) are favoured on the calcareous soils of calcium-rich pastures.

QUESTIONS

1 On average, a pair of rabbits produce a litter of six offspring four times a year. Since the offspring mature quickly, millions of rabbits could be produced within a few generations. Using the words *over-production, competition* and *variation* in your answer, explain why the world is not over-populated by rabbits.

2 Figure 17.9 shows how a mutant form of a species that possesses some advantageous characteristic could spread through a population. Imagine that, before dying, each mutant form leaves an average of two offspring as part of the next generation; each wild type leaves only one.

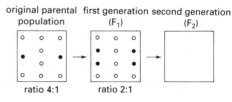

Figure 17.9

a) Which symbol represents the mutant form?
b) Draw and complete a box to represent the F$_2$. State the ratio that applies to the members of its population.
c) Continue the series of diagrams until you can state the generation in which:
(i) mutants outnumber wild type for the first time;
(ii) the ratio of wild type to mutants is 1:8.

3 Antibiotics have been used widely in agriculture to promote growth of farm animals. Suggest a long term disadvantage that could result from this practice.

4 As many as 40% of the people living in Central Africa are found to be heterozygous for the sickle cell condition. However only 4% of American Negroes (descendants of slaves from this region of Africa) suffer the sickle cell trait. Explain why.

5 The graph in figure 17.10 shows the results of a survey of lichens and melanic moths which was carried out in the vicinity of a large industrial city in the 1950s.

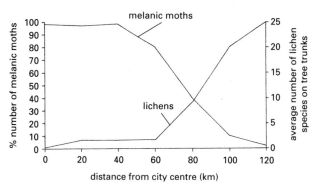

Figure 17.10

a) State the relationship that exists between the distribution of lichens and melanic moths.
b) Explain in evolutionary terms why the melanic moth graph takes the form shown in figure 17.10.
c) Predict the form that a graph of the pale-coloured peppered moth numbers would have taken if drawn using the same axes.
d) In the centre of the same city the percentage number of melanic moths decreased from 95% in 1961 to 89% in 1974. Suggest why.

6 Table 17.3 gives the key to the habitat profiles shown for two species of Viola in figure 17.11.
a) Which species is the (i) calcicole, (ii) calcifuge?
b) (i) Which type of violet could be found growing in a habitat with moist, slightly acidic soil?

moisture content of soil	very wet	wet	damp	moist	dry
acidic or calcareous (calcium-rich) soil	very acidic	slightly acidic	neutral	slightly calcareous	very calcareous

Table 17.3

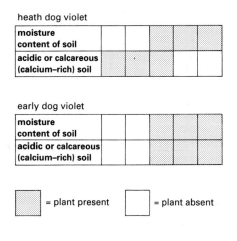

Figure 17.11

(ii) Which of these types of violet would be commonly found on very wet, very acidic soil?
c) (i) Describe TWO features of a soil on which both of these types of violet could survive.
(ii) Suggest why, in practice, they are rarely found growing together in the wild on such a soil.

DATA INTERPRETATION

For many years, rats were successfully controlled by a poison called **warfarin** which interferes with the way in which vitamin K is used in the biochemical pathway shown in figure 17.12.

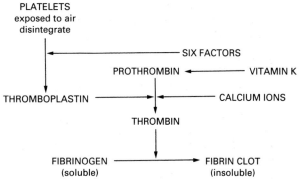

Figure 17.12

However in 1958 strains of rat appeared in Scotland that were resistant to warfarin. This resistance is an inherited characteristic as shown by the data in table 17.4.

rat's genotype	rat's phenotype
W^sW^s	sensitive to warfarin
W^sW^r	resistant to warfarin (but needs some extra vitamin K in diet)
W^rW^r	resistant to warfarin (but needs 20 times normal amount of vitamin K in diet)

Table 17.4

a) With reference to the diagram, explain why warfarin is lethal to normal rats. (1)
b) Why are people who suffer thrombosis (internal blood clotting) given small amounts of warfarin to take? (1)
c) Suggest how the resistant strain of rat arose. (1)
d) Is allele W^r dominant or recessive to allele W^s? Explain your answer. (2)
e) The resistant strain increased greatly in number over the years. Explain this success in terms of natural selection. (1)
f) If use of warfarin is continued, predict the fate of the normal wild type rat. (1)
g) Construct a hypothesis to account for the fact that most rats resistant to warfarin are heterozygotes. (1)
h) (i) Construct a diagram to show the outcome of crossing two heterozygotes.
(ii) Experts claim that the allele W^s cannot disappear completely from the rat population. Are they correct? Explain your answer. (2)

(10)

18 Selective breeding of animals and plants

Hybridisation

A **hybrid** is an individual resulting from a cross between two genetically dissimilar parents of the same species. Breeders often deliberately cross members of one variety of a species which have certain desirable features with those of another variety possessing different useful characteristics in an attempt to produce at least some hybrids which have both. A race horse breeder, for example, might try crossing a strong stallion with a fast mare.

Hybrid vigour

Hybridisation often produces some hybrids that are stronger, more fertile, or better in some other way than either parent. This is called **hybrid vigour (heterosis)**. It is especially pronounced when it involves a cross between two strains that have become homozygous as a result of many generations of inbreeding (see page 100).

Hybridisation re-establishes **heterozygosity** and poorer recessive alleles may be masked by better dominant ones as shown in the following imaginary example.

AABBccdd   ×   aabbCCDD
(true breeding   (true-breeding
strain 1)   strain 2)

AaBbCcDd
(hydrids inherit all dominant alleles for vigour and fertility, for example)

Bedding plants

Professional plant breeders maintain two different true-breeding parental lines of annual bedding plants such as marigold, pansy and snapdragon. One line might be true-breeding for flower colour (CC) and uniform growth habit

(GG), the other for large flower size (SS) and number (NN).

When these are crossed, all the F_1 hybrids possess all the desirable features and outclass both parents.

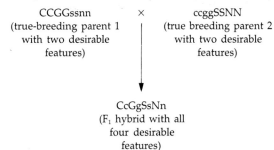

CCGGssnn   ×   ccggSSNN
(true-breeding parent 1   (true breeding parent 2
with two desirable   with two desirable
features)   features)

CcGgSsNn
(F_1 hybrid with all four desirable features)

Since the hybrids are heterozygous and therefore not true-breeding, the hybridisation process using the original parental lines must be repeated every year.

Farm animals

Hybridisation is commonly practised by farmers on their animal stocks. Table 18.1 lists some examples of traits showing hybrid vigour in cattle.

The following crosses show the outcome of

	trait	% improvement shown by hybrid
dairy cattle	birth weight	3–6
	no. live calves per calving	2
	feed conversion efficiency	3–8
	milk yield	2–10
beef cattle	birth weight	2–10
	calves weaned per cow mated	9–15
	feed conversion efficiency	1–6

Table 18.1 Hybrid vigour in cattle

hybridisation between different varieties of sheep.

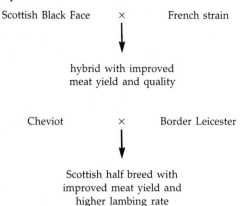

Scottish Black Face   ×   French strain

hybrid with improved meat yield and quality

Cheviot   ×   Border Leicester

Scottish half breed with improved meat yield and higher lambing rate

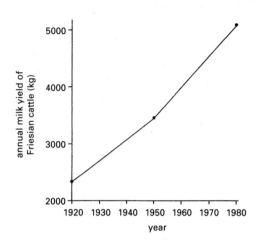

Figure 18.2   Effect of artificial selection on milk yield

Artificial selection

For thousands of years, humans have cultivated plants and domesticated animals for their own benefit. Breeders select those individuals which have characteristics nearest to what they want and use them to breed the next generation. They prevent those with undesirable characteristics from breeding.

This process is called **artificial selection**. It leads to continuing change in domestic varieties (see figure 18.1). The breeder does not, however, create the new variation. The variation already existed amongst the members of the species. Breeders simply accumulate certain desirable parts of it within pedigree strains.

From wild cabbage (*Brassica oleracea*), breeders have selectively bred modern cabbage, cauliflower, broccoli, Brussel sprouts and kale.

Certain characteristics of domesticated animals have been enhanced through selective breeding. Friesian cattle have been selected for milk yield (see figure 18.2). Similarly meat production in Aberdeen Angus cattle has increased over the years.

By employing artificial selection, humans often direct another species along an evolutionary path that it would not otherwise have taken. Slow moving cattle with heavy udders, for example, would be easy prey to predators and would not enjoy a selective advantage in the wild.

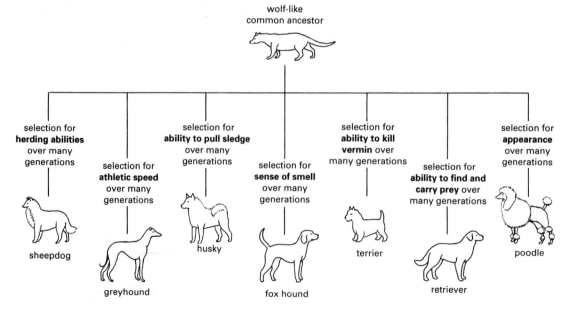

Figure 18.1   Artificial selection of pedigree dog strains

Imported species

Scotland's economy is further enriched by the introduction of species from other countries. Certain **imported species** are selected for their ability to thrive in the Scottish climate and offer some important economic benefit.

Sitka spruce

This native of North America grows well on cold, damp, inhospitable land. It provides a valuable source of light, strong, fine-grained timber and its growth rate is much faster than native conifers such as Scots pine.

Rainbow trout

These fish from North America are used to stock fish farms where they are maintained on artificial diets until their size is optimal and the market price is right. They are less demanding to farm than their relative, the Atlantic salmon.

Kashmir goats

Animals, semen and embryos are imported from China and Russia. The wool of these goats is of great economic value.

Rape

In recent years British farmers have been encouraged to produce crops of the rape plant (a member of the cabbage family). This plant is a traditional crop of many European and Asian countries. By 1983, it had become the most widely grown non-cereal crop in Britain. It is of economic value because its seeds are rich in edible oil. In addition its leaves provide food for farm animals.

QUESTIONS

1 Table 18.2 refers to a breeding experiment involving cattle. What can be concluded from the data?

	genotype	weight gained by phenotype per day (kg)
parent 1	AAbb	0.82
parent 2	aaBB	0.82
F$_1$ generation	AaBb	1.00

Table 18.2

2 The cultivated variety of tomato plant is susceptible to attack by eelworm. A wild variety of tomato plant which bears fruit of poor quality is resistant to eelworm attack.
a) Describe how scientists could attempt to develop a variety of tomato plant better than either of these types.
b) Suggest why their attempts might not be immediately successful.

3 a) What is meant by the term **artificial selection**?
b) What is the purpose of artificial selection?

4 A breed of pig has been produced which reaches bacon weight in 150 days instead of the normal 185 days. In addition, this is achieved on 20% less food.

Can this improvement process be continued indefinitely? Explain your answer.

19 Loss of genetic diversity

The earth supports an extraordinary variety of interdependent life forms upon which natural selection has been acting for millions of years. Each member of a species possesses hundreds or even thousands of genes. Since two or more alleles exist for most genes, the number of genetic permutations possible amongst the members of a species is enormous.

This potential for **genetic diversity** amongst its living inhabitants, is one of our planet's most valuable resources.

Loss of diversity by inbreeding

Homo sapiens is the first life form to have been able to deliberately alter the evolutionary process. This has been done by practising artificial selection (see page 98) on a chosen few species of plants and animals. Once a desirable strain is established, farmers often try to produce a uniform crop or herd with the desirable characteristics by employing **inbreeding**. This is the process by which an individual is bred with its close relatives and is prevented from mating at random. (Self-pollination in plants is the most intense form of inbreeding.)

Continuous inbreeding results in *loss* of genetic diversity. Table 19.1 shows how the inbreeding of heterozygotes results in heterozygosity being halved at each generation. Eventually the population would become (virtually) homozygous at that locus producing two pure lines AA and aa. In reality many gene loci are involved and there are many different ways in which lines may become homozygous e.g.

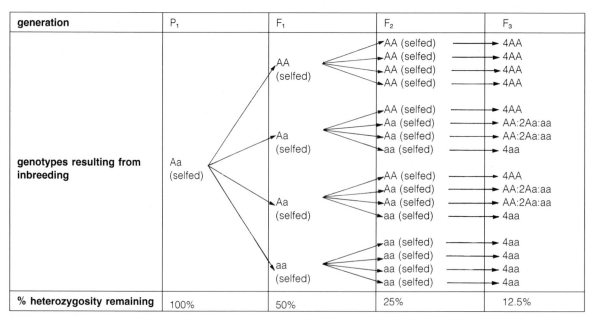

Table 19.1   Effect of continuous inbreeding

AAbbCC . . . etc, aabbCC . . . etc, AABBcc . . . etc and so on.

Inbreeding therefore increases the chance of individuals arising that are double recessive for some harmful allele (e.g. poor resistance to disease). Such individuals are biologically and economically inferior and are said to show **inbreeding depression**. It is for this reason that inbreeding is not carried out indefinitely. Instead, new alleles are introduced by **hybridisation** (see page 97).

Loss of diversity in a small population

In an attempt to satisfy the demands of the ever increasing population, humans chop down, plough up, dam and often pollute natural habitats. In doing so they threaten the very existence of many of the species living there. An area of tropical rain forest the size of Scotland is at present being destroyed annually. If this continues, many species already in danger will be lost forever.

Even when a species is recognised as being endangered, it is often too late to restore it to its natural diverse state because already its gene pool (see page 110) has been greatly depleted. Take for example a species of a million individuals reduced by habitat destruction to say 10 000. Certain alleles may have already been completely lost. The smaller this population becomes, the greater the chance that all the individuals possessing a certain allele will perish.

It is for this reason that *irreversible loss* of genetic diversity is closely associated with small population size of endangered species.

Need for conservation of species

Humans are able to destroy in a decade what took millions of years to evolve. At last we have begun to appreciate that nature is not inexhaustible. It is essential for our very survival that other species are **conserved** for their genetic potential to provide future food, fuel, medicines and raw materials. In addition we must have a supply of **alternative alleles** for introducing into domesticated varieties to make them more vigorous and more resistant to disease in years to come.

Protecting genetic diversity

On-site protection involves identifying an ecosystem containing a valuable gene pool and trying to maintain it in its natural state. This could mean making it into a game reserve or national park.

Figure 19.1 shows regions of the world called **centres of diversity** from where modern day food plants originated. Since these regions still possess **wild relatives** of our food plants, they are prime targets for conservation.

Off-site protection means identifying a potentially valuable species and maintaining it in an artificial environment.

This may take the form of a **rare breed farm** where varieties of farm animal (or their ancestors) which have otherwise become extinct are bred to provide a reservoir of alternative genes.

In addition **gene (germ cell) banks** allow seeds and germ plasm (sex cells) to be stored in a dormant state for long periods if kept at very low

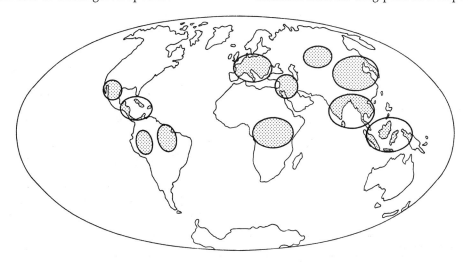

Figure 19.1   Centres of genetic diversity

temperature and humidity. By this means, many thousands of species including wild varieties of crop plants can be conserved for future use.

Table 19.2 lists some of the wild relatives of the domesticated tomato plant (*Lycopersicon esculentum*) which are held in a germ plasm collection. These possess many useful characteristics under genetic control which are already being introduced into the domestic tomato to produce improved varieties.

Although the earth possesses at least 75 000 species of edible plant, man depends on only 30 of these to produce 95% of the world's food (see figure 19.2). Experts forecast that global temperatures will rise by 3°C in the next 50 years as a result of the 'Greenhouse Effect'. If this occurs many of today's crops may fail in the warmer climate. It is for this reason that new gene banks are being developed. However it is not possible to store every variety of every species. Priority is being given to wild drought-resistant relatives of cereals and legumes in the hope that new varieties able to survive the Greenhouse Effect can be developed in the future.

wild relative of tomato	native habitat	useful features controlled by genes
Lycopersicon cheesmanii	Galapagos Islands	ability to absorb water from sea water, possession of jointless fruit stalks allowing easy mechanical harvesting
L. chilense	Chile and Peru	drought resistance
L. esculentum cerasiforme	many tropical and sub-tropical countries	ability to tolerate high temperature and humidity
L. hirsutum	Ecuador and Peru	resistance to various pests and viruses
L. pennellii	Peru	resistance to drought, high sugar and vitamin A and C content
L. peruvianum	Peru	resistance to various pests, high vitamin C content
L. pimpinellifolium	Peru	resistance to many diseases, improved fruit quality, high vitamin C content

Table 19.2 Wild relatives of tomato plant

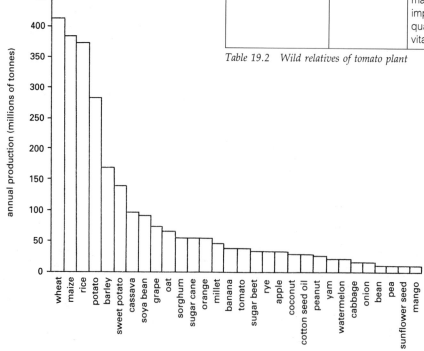

Figure 19.2 Reliance on 30 plant species

QUESTIONS

1 State ONE advantage and ONE disadvantage of producing a crop which is uniform and true-breeding.

2 a) Draw TWO conclusions from the graph in figure 19.3.

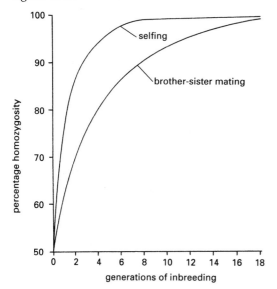

Figure 19.3

b) Most religions and countries forbid marriages between close relatives. What is the advantage of this policy from a genetic point of view?

3 *'A small population of an endangered species is more likely to suffer loss of genetic diversity than a large population of a common species.'*
Do you agree or disagree with this statement? Explain your answer.

4 a) Explain why the conservation of other species is directly related to the future survival of humankind on earth.
b) What is the difference between on-site and off-site protection of species?
c) What is a germ cell bank and why is it important?

DATA INTERPRETATION

The data in tables 19.3 and 19.4 refers to breeding experiments using maize (*Zea mays*), see figure 19.4.
a) Identify the generation formed by hybridisation. (1)
b) State THREE pieces of information which provide evidence of hybrid vigour. (3)
c) Which of the following was a true-breeding inbred line:
P_1, F_1, F_3, F_5? (1)
d) Using an example from the data to illustrate your answer, explain what is meant by the term **inbreeding depression**. (2)
e) In 1933, American farmers began planting maize grains produced by hybridisation. This practice has continued over the years and has brought about the changes shown in table 19.4.
(i) State TWO ways in which farmers have benefited by using hybrid maize. (2)

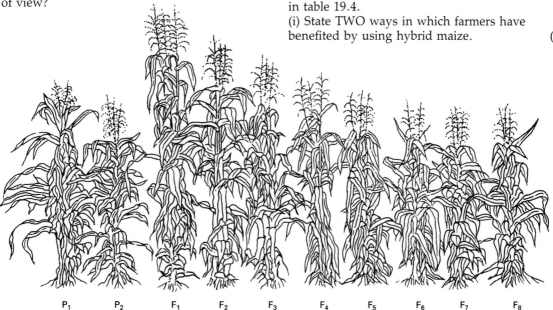

P_1 P_2 F_1 F_2 F_3 F_4 F_5 F_6 F_7 F_8

Figure 19.4

	parents		successive generations (resulting from repeated self-fertilisation)							
	P_1	P_2	F_1	F_2	F_3	F_4	F_5	F_6	F_7	F_8
number of generations selfed	17	16	0	1	2	3	4	5	6	7
average ear length (mm)	84	107	162	141	147	121	94	99	110	107
average yield (metric tonnes/hectare)	2.0	2.1	4.3	4.0	3.4	3.1	2.8	2.6	2.5	2.3

Table 19.3

year	millions of hectares used for maize planting	average yield (metric tonnes/hectare)
1932	46	2.4
1946	36	3.4
1957	30	4.1

Table 19.4

(ii) Explain why farmers must buy expensive hybrid grain every year from supply houses to sow their maize crop instead of simply using grain kept back from the previous year's crop. (1)

(10)

20 Genetic engineering

Chromosome mapping

By doing test crosses and calculating the cross-over values (based on the frequency of recombination) between linked genes (see page 74), geneticists have been able to build up comprehensive **chromosome maps** of organisms such as maize and fruit fly (see figure 20.1).

Such a linear arrangement of genes on a chromosome is based on indirect evidence since genes are too tiny to be seen. However in some animals, unusual **giant chromosomes** are found to exist. The chromosomes in the salivary gland cells of fruit fly larvae, for example, are ten times longer and a hundred times thicker than normal. (This is because homologous partners remain together and replicate many times without separating. They fail to shorten in the usual way and only certain parts become coiled.)

Some parts of these chromosomes are found to take up more stain than others producing distinct **bands** which vary in thickness (see figure 20.2). The particular pattern of banding is a **constant characteristic** of each type of chromosome.

Locating a gene on a chromosome

It is possible to relate the location of individual genes to particular bands on chromosomes. The gene for red/white eye colour in fruit fly is sex-linked (see page 78). To find the location of this gene on the X chromosome, red-eyed males ($X^R Y$) are first irradiated to induce mutations. They are then crossed with white-eyed females ($X^r X^r$). Almost all of the females produced are found to be red-eyed as expected. However a few are white-eyed because they have received an X chromosome which has suffered a **deletion** at the locus of the R allele (see figure 20.3).

If the X chromosomes from the salivary glands

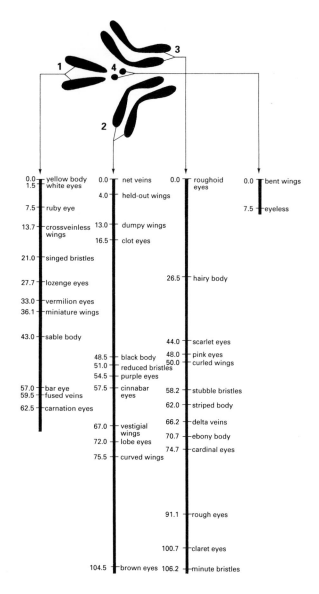

Figure 20.1   Chromosome maps (only a few genes shown)

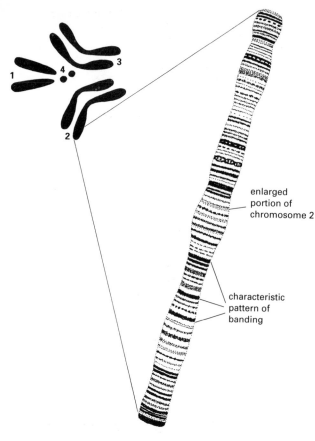

Figure 20.2   Pattern of banding on chromosome

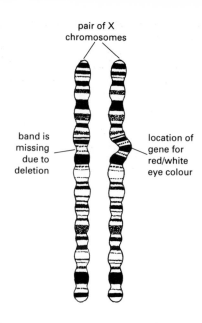

Figure 20.4   Location of deleted gene on X chromosome

Other mutations also lead to **gross chromosomal abnormalities**. For example, **inversion** causes a section of a chromosome to become reversed (see page 84). Homologous chromosomes are only able to pair up by forming a **loop** as shown in figure 20.5. If this structural abnormality can be connected with unusual results in breeding experiments, then it must be the site of the affected gene(s).

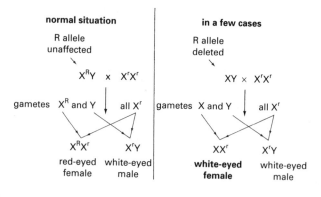

Figure 20.3   Crosses to locate a deleted gene

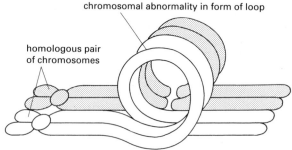

Figure 20.5   Gene location aided by chromosomal abnormality

of larvae produced by these white-eyed females are examined, some of them are found to be unusual (see figure 20.4).

Instead of matching one another band for band, one X chromosome is found to have a band missing. It is therefore concluded that this is the location of the gene for red/white eye colour.

Mapping the human genome

Since human beings cannot be subjected to breeding experiments, knowledge about the human genome is fairly limited. In the past it has depended largely upon studies of family pedigrees. Genes responsible for sex-linked diseases and disorders were known to be located on the X chromosome but this did not help to mark genes located on autosomes.

However, thanks to molecular biology and computing, human chromosomes are at last being mapped. It is estimated that each human being possesses 50 000–100 000 genes. It is possible that the complete **base pair** sequence which makes up the human genetic code, the '**handbook of man**', will be known within 20 years or less. However there remains the problem of ethics. Worldwide agreement is needed to control the uses to which this information may be put in the future.

Altering the genetic information

Recombinant DNA technology (**genetic engineering**) involves the transfer of genes from one organism (e.g. man) to another organism (e.g. bacterium). The naturally occurring mechanism of genetic recombination described on page 69 takes place within only one species. Genetic engineering on the other hand provides man with unlimited opportunities to create new combinations of genes from more than one species which would not occur naturally.

Genetic engineers are able to select a particular gene for a desirable characteristic (e.g. gene for human insulin), splice it into the DNA of a **vector** (e.g. **plasmid** from a bacterial cell) and insert the vector into the host cell (e.g. bacterium such as *Escherichia coli*). When the reprogrammed host cell is propagated, it expresses the 'foreign' gene and produces the product (e.g. insulin).

The production of recombinant DNA plasmids involves the use of two special enzymes as shown in figure 20.6. **Endonuclease** cleaves open the plasmid and cuts out the required DNA fragment so that the ends are complementary. **Ligase** seals the two together.

Future applications

Thale cress (scientific name, *Arabidopsis*) is a garden weed closely related to rape and cabbage plants. Scientists are at present analysing all of its genes in order to identify the ones responsible for *growth regulation, resistance to fungal attack* and *oil quality* of seeds. They hope to transfer these genes to crop plants thereby increasing yield with decreased use of fertilisers and pesticides.

Already some animal genes have been successfully inserted into plants. It is therefore possible that in the future some farmers may

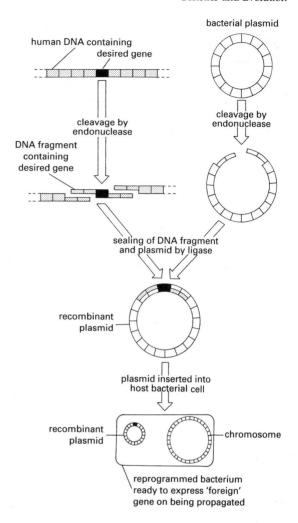

Figure 20.6 Genetic engineering

grow crops for their insulin, human growth factor or blood-clotting agents!

DNA technology is also being applied to yeasts and plant and animal cell cultures. It allows scientists to develop organisms with new properties which could never be produced by selective breeding.

Production of new crop plants by somatic fusion

Two different species cannot interbreed successfully. At best, a cross between them produces a sterile hybrid. However new techniques are enabling scientists to overcome this problem.

Unspecialised cells of two different plant species are selected and their cell walls digested away using the enzyme cellulase. This leaves structures called **protoplasts**, each consisting of living cell contents (nucleus and cytoplasm) surrounded by the cell membrane.

In some cases isolated protoplasts from two different species will fuse to form a **hybrid protoplast**. This process is called **somatic fusion**. The hybrid is induced to form a cell wall and to divide into an undifferentiated cell mass (**callus**). In the presence of hormones, calluses develop into hybrid plants containing a mixture of the parents' genetic traits.

Resistance to potato leaf roll
Potato leaf roll is a disease caused by a virus which is spread by aphids. It can severely reduce the yield of a crop of the potato plant (*Solanum tuberosum*). *Solanum brevidens* is a wild non-tuber-bearing species of potato plant which is *resistant* to potato leaf roll virus.

Scientists have overcome the lack of sexual compatibility between these two species of plant by uniting them by somatic fusion. This has resulted in the production of a new variety of potato plant which is tuber-bearing *and* resistant to potato leaf roll virus.

Scientists are also trying to unite wheat with pea plants by somatic fusion. If successful, the somatic cell hybrid might combine the food-producing features of wheat with the ability of pea plants to form root nodules containing nitrogen-fixing bacteria.

Already hybridisation between potato and tomato plants has been successfully performed by somatic fusion. However none of the hybrids produced so far is able to make both potato tubers and tomato fruit on one plant.

QUESTIONS

1 Figure 20.7 refers to the fruit fly where + = allele for normal eye, and B=allele for bar eye.
   a) Is the allele for bar eye dominant or recessive to normal eye? Explain your answer.
   b) What type of chromosome mutation caused the chromosomal abnormality?
   c) On which numbered section of the X chromosome is the gene for bar eye located?
2 Briefly describe the role played by the enzymes endonuclease and ligase during the process of genetic engineering.
3 Production of the human growth hormone **somatotrophin** is controlled by a gene. Lack of somatotrophin leads to dwarfism in humans. This problem has now been solved by the production of somatotrophin by genetic engineering. Give a step by step account of how this could have been achieved.
4 Briefly explain why some scientists are predicting that certain crop plants could

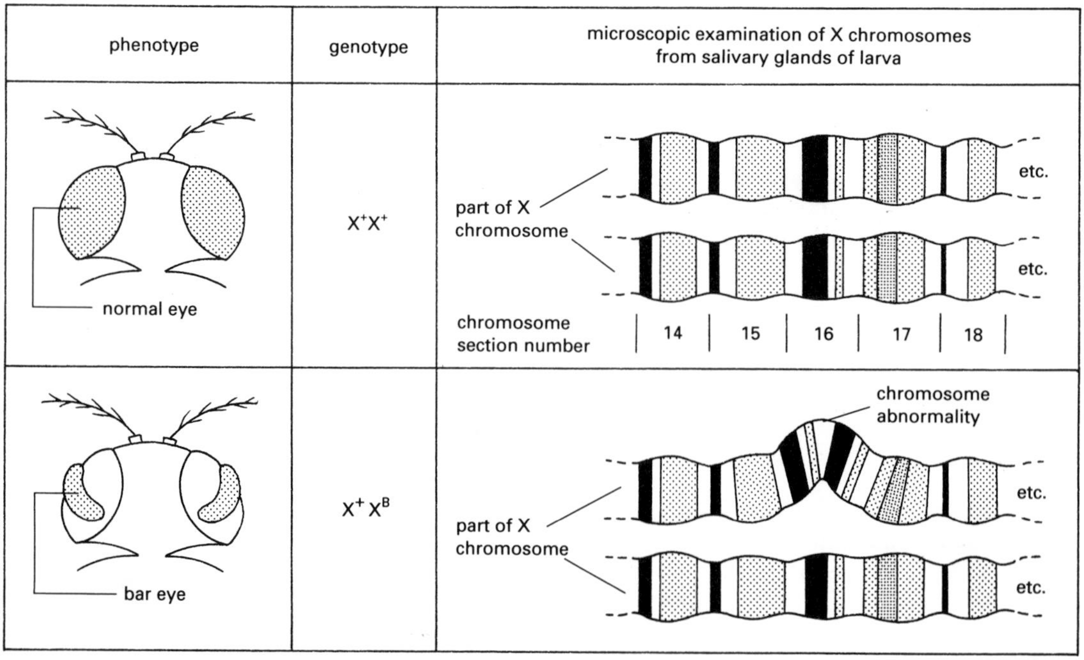

phenotype	genotype	microscopic examination of X chromosomes from salivary glands of larva
normal eye	X^+X^+	part of X chromosome / chromosome section number 14 15 16 17 18 etc.
bar eye	X^+X^B	part of X chromosome / chromosome abnormality etc.

Figure 20.7

become the pharmaceutical factories of tomorrow.

5 In nature, hybridisation between sugar beet and cabbage plants never occurs because they are unrelated species. Suggest a method by which plant scientists could attempt to combine the genetic material of the two in the laboratory to produce a useful hybrid.

What you should know (CHAPTERS 17–20)

1 **Natural selection** favours those members of a population best suited to an environment.
2 Rare mutant forms sometimes enjoy a **selective advantage** if some biotic or abiotic factor brings about a change in the environment.
3 The use of **hybridisation** in plant and animal breeding often produces offspring which show **hybrid vigour**.
4 Breeders use **artificial selection** to selectively breed useful organisms.
5 **Loss** of **genetic diversity** is associated with **inbreeding** of domesticated plants and animals. It may also occur in **small populations** of endangered species. Important species are therefore often **conserved** in rare breed farms and gene banks.
6 The **location** of genes on chromosomes can be **mapped**.
7 Using **recombinant DNA technology**, scientists are able to combine genetic material from two very different species and produce organisms which would never otherwise arise.

21 Speciation

Gene pool

The total of all the different genes in a population is known as the **gene pool**. The frequency of occurrence of an allele of a gene in a population (relative to all the other alleles at the same locus) is known as the **gene frequency**.

As long as the population is **large**, and mating is at **random**, and none of the conditions listed below occurs, then the gene frequencies in the population remain constant from generation to generation. Such stability is called **genetic equilibrium**.

Alteration to gene pool (gene frequencies)

The process of evolution is dependent upon changes occurring in a gene pool. These changes are caused by the following six factors.

Mutation

New alleles appear and are reshuffled by independent assortment of chromosomes and crossing over during meiosis.

Natural selection

Particular alleles are favoured and their frequency increases.

Gene migration (immigration and emigration)

New alleles are introduced when immigrants arrived from another population possessing a different gene pool.

Non-random mating

Continuous inbreeding leads to a decrease in heterozygosity, (see page 100).

Genetic drift

When a small group of organisms becomes isolated from the rest of the population, the splinter group often does not possess the complete range of gene alleles typical of that species. After several generations, the isolated group becomes distinctive since the frequencies of certain alleles have changed.

For example, it is thought that North America was first populated by a small group of Asians who migrated across the ice packs of Bering Strait and became isolated from the rest of the Mongoloid race. **Genetic drift** accounts for the differences in the percentages of population possessing certain blood groups. This is shown in table 21.1.

	% population with blood group			
people	A	B	AB	0
Chinese	31	28	7	34
Sioux American Indians	7	2	0	91

Table 21.1   Genetic drift

Chance

In a small population there is also the risk that some fairly rare allele will be completely lost by chance (e.g. forest fire).

Species

A **species** consists of a group of organisms that: *share common anatomical and physiological characteristics and have the same chromosome complement and gene pool enabling them to interbreed and form* **fertile offspring**.

They are reproductively isolated from other such groups (different species).

It is thought that there are about 5–10 million species on earth at present. However species are not constant immutable units. Their number and kinds are always changing. At any given moment some species will be enjoying a stable relationship with the environment, some will be moving towards extinction and others will be undergoing speciation.

Extinction

Ever since life evolved on earth, new species better suited to the environment have appeared. Older less successful forms have died out. Fossil evidence shows that life on Earth has been punctuated by several waves of **mass extinction**. These are closely related to changes in global climate that have occurred. For example, during an Ice Age, sea levels drop and vast areas of frozen land appear. As a result, many animals which are native to warm, sheltered, marine waters and estuaries perish.

When the environment becomes disrupted, species that were previously successful may suddenly find themselves at a major disadvantage and become extinct. Other species which had been enjoying minimum success in a hostile environment may find themselves well suited to the new environment and evolve rapidly.

Mammals, for example, existed as a relatively insignificant group during the age of the dinosaurs (see page 118) and only became the earth's dominant life form when some major change in the environment caused the dinosaurs to die out.

Man's activities are causing the current wave of extinction to run at about **four hundred times** the natural rate. Since the year 1600, hundreds of birds and mammals have been wiped out by **over-hunting** and by **habitat destruction**. Table

region	animal	extent of danger
Africa	black rhinoceros	critical
	mountain gorilla	high
	elephant	moderate
North America	Californian condor	critical
	grizzly bear	moderate
Asia	Siberian tiger	critical
	panda	high
	snow leopard	high
World's oceans	blue whale	high

Table 21.2 Endangered species

21.2 shows a few examples of the many species now threatened with extinction.

In addition to these 'high profile' animals, a much larger number of plants and invertebrate species (e.g. insects) are in critical danger as man continues to destroy their ecosystems. At the present rate, it is estimated that man will have wiped out one million species by the year 2000.

If this process of destruction continues, the future will be bleak. Species necessary for mankind's survival will be lost (see page 101) and many ecological niches will become populated by opportunist species such as cockroaches, rats and weed plants.

Speciation

Speciation is the **formation** of new species (usually many from a few). **Evolution** is the **mechanism** by which speciation is brought about. It involves changes in genotype (and phenotype) of a population. These changes are adaptive, making the organism better at exploiting the environment.

A simplified version of speciation is shown in figure 21.1.

1 The members of a large population of a species occupy an environment. They share the same gene pool and interbreed freely.

2 The population becomes split into two completely isolated sub-populations by a **barrier** which prevents interbreeding and gene exchange. The barrier may be:

geographical (see figure 21.2);
ecological caused by changes in temperature, humidity, pH etc.;
reproductive.
a) different populations become sexually receptive at different times of the year,
b) members of different populations are not attracted to one another or fail to be stimulated by one another's courtship behaviour,
c) pollination mechanisms fail,
d) cross-fertilisation is prevented by physical non-correspondence of sex organs,
e) gametes fail to fuse,
f) offspring are sterile.

3 **Mutations** occur at random in each sub-population, giving rise to **new variation** within each group but not shared by both groups.
4 The selection pressures acting on each sub-population are different depending on local conditions such as climate, predators, disease etc. **Natural selection** affects each sub-group

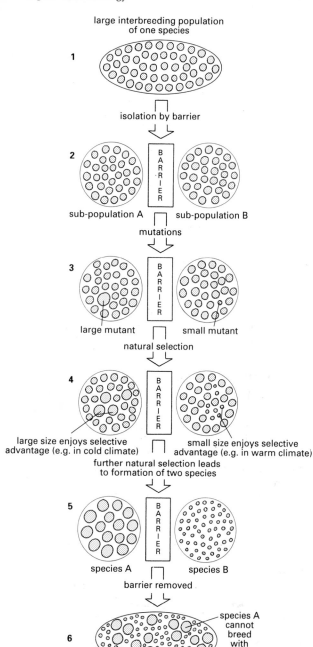

Figure 21.1   Speciation

in a different way by favouring those alleles which make the members of that sub-population best at exploiting their environment. (If one of the original sub-populations is small, genetic drift may also play a role in the alteration of gene frequencies.)

5 Over a very, very long period of time, stages

3 and 4 cause the two gene pools to become so altered that the groups become genetically distinct and **isolated**.

6 If the barrier is removed, they are no longer able to interbreed (since their chromosomes cannot make homologous pairs). **Speciation** has occurred and two separate distinct species have evolved.

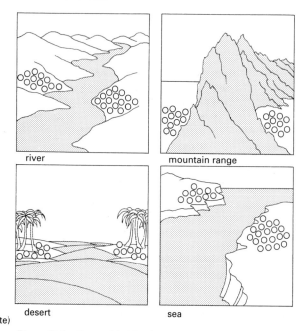

Figure 21.2   Geographical barriers

Island populations

When a particular type of organism is found only in a certain region (e.g. an island), it is said to be **endemic** to the region. Endemic species of plants and animals are found to occur on some of the Scottish Islands.

Sorbus

Sorbus aucuparia is the scientific name given to the mountain ash (rowan) tree which is commonly found throughout Europe, especially on rocky mountainous soil but not on clay or limestone. *Sorbus rupicola* is a shrub (or rarely a small tree) found on crags amongst rocks especially on limestone soil.

Sorbus arransensis (Arran whitebeam) is a small slender tree endemic to the island of Arran where it grows on steep granite stream banks. It is believed to have arisen by hybridisation between *S. aucuparia* and *S. rupicola*. By growing in habitats intermediate to those favoured by its

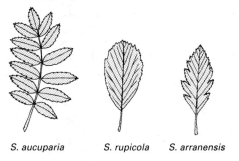

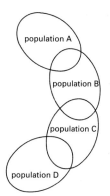

populations interbred	result
A × B	+
B × C	+
C × D	+
A × C	−
A × D	−
B × D	−

+ = successful interbreeding
− = unsuccessful interbreeding

Figure 21.5   Interruption of gene flow

Figure 21.3   Leaves of three species of Sorbus

parents, and lacking competitors found on the mainland, *Sorbus arransensis* has taken its own course of evolution and is different from both of its parents (see figure 21.3).

European wren

Figure 21.4 refers to four subspecies of the European wren, *Troglodytes troglodytes*. Geographical isolation from the original mainland population, and from one another, has caused each island species to take its own course of evolution.

Although the populations have not been isolated for long enough to allow distinct species to arise, it is interesting to note that the subspecies on St. Kilda, which is furthest from the mainland, is already the most different.

Similarly four different subspecies of Orkney vole are found on four different islands in the Orkney group. (See also Data Interpretation on page 119.)

Interruption of gene flow

An environment may be occupied by several populations of an organism which vary in their ability to interbreed as shown in figure 21.5. At

present, the four populations belong to the same species since genes can flow from A to D via B and C. However if population B or C disappeared then gene flow would be interrupted and A and D would become genetically isolated and form **two distinct species**.

Continental distribution of organisms

The world's land mass can be divided into several distinct regions according to the types of animals that each possesses. Figure 21.6 shows only a small sample of the **fauna** (native animals) of each region.

When the full range for each region is considered in detail, it is found that the mammals of regions **1** and **2** are very similar to one another, those of regions **3, 4** and **5** fairly different, and those of **6** very different from all of the others.

Such **continental distribution** of mammals is thought to have occurred as a result of **continental drift**. The world's land masses are thought to have been joined together many millions of years ago (see figure 21.7) and then to have gradually drifted apart.

It is thought that ancestral stocks of animals originated in the northern hemisphere and that some of these migrated to southern continents still connected by **land bridges**. The populations later became cut off from one another by geographical **barriers** such as water, mountains and deserts.

Regions **1** and **2** with their very similar animals are thought to have become separated only relatively recently. Region **6** with its very different animals has been isolated from the other land masses for the longest time.

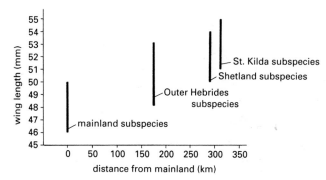

Figure 21.4   Island subspecies of wren

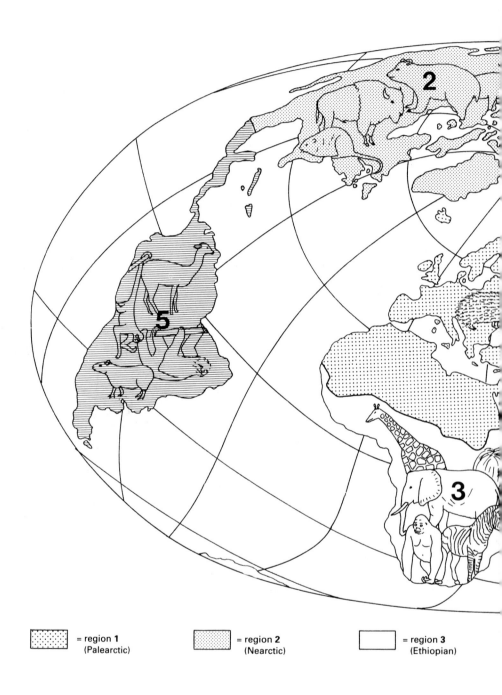

= region **1**
(Palearctic)

= region **2**
(Nearctic)

= region **3**
(Ethiopian)

Figure 21.6   Continental distribution of mammals

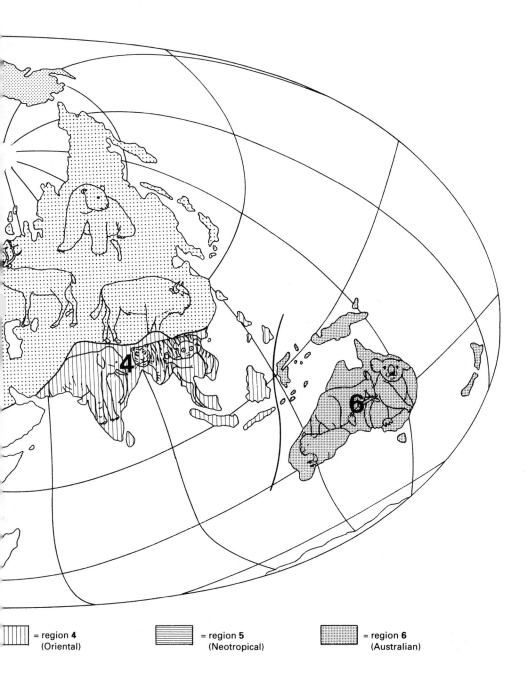

| | = region **4**
(Oriental) | | = region **5**
(Neotropical) | | = region **6**
(Australian) |

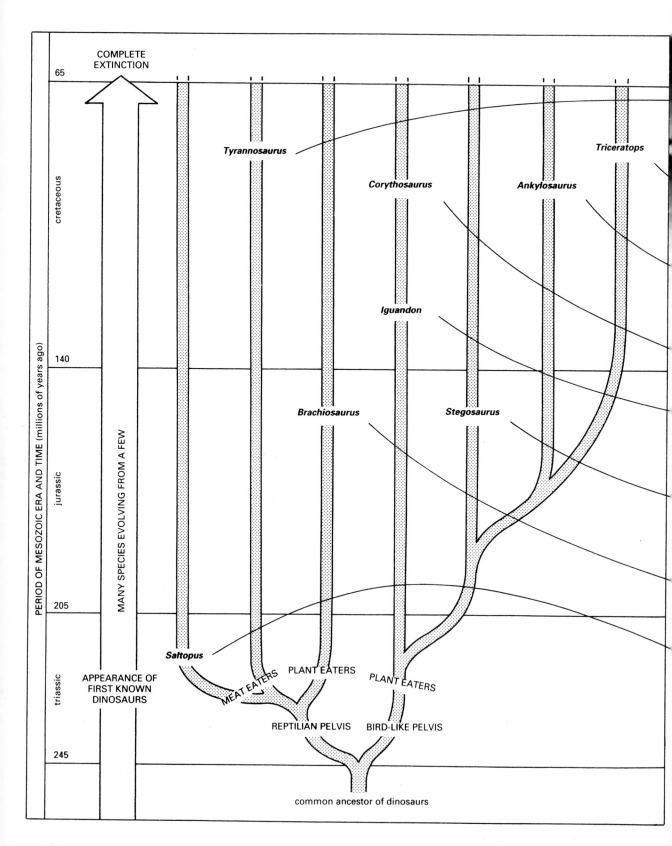

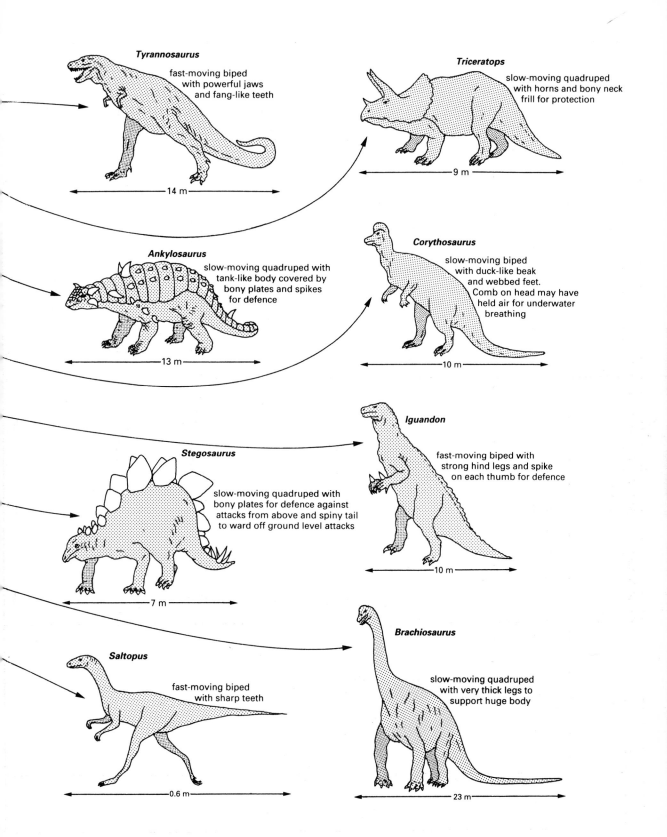

Tyrannosaurus
fast-moving biped with powerful jaws and fang-like teeth
14 m

Triceratops
slow-moving quadruped with horns and bony neck frill for protection
9 m

Ankylosaurus
slow-moving quadruped with tank-like body covered by bony plates and spikes for defence
13 m

Corythosaurus
slow-moving biped with duck-like beak and webbed feet. Comb on head may have held air for underwater breathing
10 m

Stegosaurus
slow-moving quadruped with bony plates for defence against attacks from above and spiny tail to ward off ground level attacks
7 m

Iguandon
fast-moving biped with strong hind legs and spike on each thumb for defence
10 m

Saltopus
fast-moving biped with sharp teeth
0.6 m

Brachiosaurus
slow-moving quadruped with very thick legs to support huge body
23 m

Figure 21.8   Evolutionary tree of dinosaurs

117

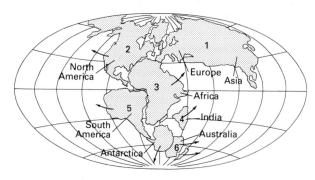

Figure 21.7 Continental drift

Conservation of the marsupials

Whereas the most advanced group of mammals, the **placentals**, retain their young in the uterus until a late stage of development, the more primitive **marsupials** bear their young at a very early stage and rear them in a pouch.

Since there are no native placentals in Australia, it is thought that only marsupials reached region **6** (Australia) via region **5** (South America) and Antarctica before separation of the land masses. When the more successful placentals arrived later in region **5**, only a few marsupials survived the competition.

However in Australia, in the absence of placentals, the marsupials were able to diversify and fill every available **ecological niche** (see page 121). In this case, the isolating barrier (water) led therefore to the conservation of the rich assortment of Australian marsupials that survive to this day.

Origin and evolution of species

Dinosaurs

Fossils provide evidence for the existence on earth of at least 150 different species of **dinosaur** during the Mesozoic Era from 225 to 65 million years ago. Dinosaurs were advanced animals and there had been a long history of life on earth before they evolved.

Since the rocks containing the fossils are of different ages, this shows that all species of dinosaur did not live at the same time but gradually evolved over a period of 160 million years.

In addition to estimating when a dinosaur lived, experts can work out which dinosaurs were related by identifying structural characteristics that they had in common. From this information it is possible to build up an **evolutionary tree** (figure 21.8). Each branch represents a related group of dinosaurs. A very few examples have been highlighted in the diagram.

Upper Triassic rocks from Northern Scotland show that *Saltopus*, one of the earliest dinosaurs, was no bigger than a turkey. However, during the Jurassic period, many enormous dinosaurs evolved all over the world. During the Cretaceous period, flowering plants appeared and a wide variety of plant-eating dinosaurs also evolved, many possessing 'bizarre' structural adaptations to aid survival.

Although still the subject of heated debate, the reason for the 'sudden' extinction of the dinosaurs is unknown. It is possible that they were unable to adapt to changes in world climate.

Geum

Geum rivale (water avens) and *Geum urbanum* (wood avens) are two closely related species which differ in structure as shown in figure 21.9. Although both species commonly occur over much of Europe (see figure 21.10), they are usually separated by ecological differences. *G. rivale* favours very damp shady habitats in upland regions; *G. urbanum* is found in hedgerows and marginal woodland in lowland regions affected by man's activities. In addition, their different flower types tend to segregate their insect pollinators.

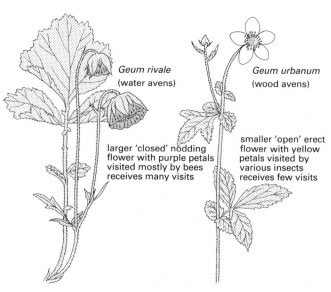

Figure 21.9 Comparison of two species of Geum

118

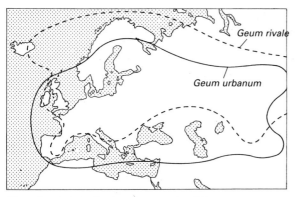

Figure 21.10 *Distribution of two species of* Geum

However, when the two do meet in damp woodland, they form fertile hybrids! They are therefore very closely related and, although undergoing speciation, are not yet, strictly speaking, two separate species.

It is thought that *G. urbanum* originally evolved either from *G. rivale* or a common ancestor in south-east Europe as a result of geographical isolation. The fact that *G. rivale* and *G. urbanum* are no longer isolated is probably due to man's activities (such as hedgerow planting) which have favoured the spread of *G. urbanum* throughout much of Europe.

By gradually breaking down the ecological isolation that separates *G. rivale* from *G. urbanum*, man is unwittingly reducing the chance of speciation occurring. In the future, increased hybridisation between the two 'species' may reunite them. However on islands such as Iceland which lacks *G. urbanum*, *G. rivale* will take its own course of evolution.

QUESTIONS

1 When a horse is crossed with a donkey, the result is a sterile animal called a mule. Do a mule's parents belong to the same species? Explain your answer.

2 The sex of certain reptiles is known to be determined by the temperature at which their eggs are incubated. In the alligator, 30°C produces females and 33°C males. It is possible that this was also true of the dinosaurs. It is known that the end of the age of the dinosaurs coincided with a lowering of the Earth's temperature.

Based on the above information, construct a hypothesis to account for the extinction of the dinosaurs.

3 Using the information in figure 21.7, construct a bar graph to compare the relative sizes of the eight dinosaurs shown.

4 a) Arrange the following stages into the correct order in which they would occur during speciation:
(i) isolation of gene pools;
(ii) formation of new species;
(iii) mutation;
(iv) occupation of territory by one species with one gene pool;
(v) natural selection.
b) Name THREE types of barrier that could bring about isolation of gene pools.

5 *Cerastium arcticum* is the scientific name for mouse-ear chickweed which is a common plant on the Scottish mainland. *Cerastium arcticum edmondstonii* is a subspecies of mouse-ear chickweed endemic to the Shetland Isles and found nowhere else in the world.
a) Suggest how this subspecies originated.
b) Under what conditions may it evolve into a separate species of *Cerastium*?

6 Give a possible explanation for each of the following observations:
a) When Australia was first discovered by European explorers in the seventeenth century, it possessed a rich variety of marsupial mammals but the only placentals were man (Aborigine) and the semi-domesticated dog (dingo).
b) When New Zealand was discovered at about the same time, it possessed no marsupials. The only mammals were man (Maori) and two species of bat.

DATA INTERPRETATION

The data in table 21.3 and figure 21.11 refer to the three subspecies of the fieldmouse (*Apodemus sylvaticus*).

Some scientists suggest that the fieldmouse became established on the Shetland Isles and Fair Isle millions of years ago while these islands were attached to the mainland. Following separation, each island population then took its own course of evolution.

However this theory is now disputed by the work of Professor R. J. Berry. From detailed studies of the skeletons of different subspecies of *Apodemus*, this scientist has devised a way of measuring how close or distant the relationship is between two subspecies. This is expressed as

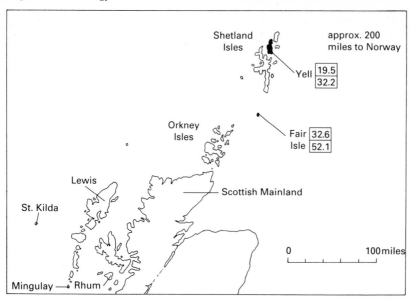

Figure 21.11

genus	scientific name species subspecies		location	average body length (mm)	average tail length (mm)	ventral colour	pectoral spot
Apodemus	*sylvaticus*	*sylvaticus*	Scottish mainland	92	83	dull white with slate-grey throat and belly	buff coloured area between front legs
Apodemus	*sylvaticus*	*granti*	Yell (Shetlands)	101	88	dull bluish-white	small spot between front legs
Apodemus	*sylvaticus*	*fridariensis*	Fair Isle	113	99	dull bluish-white	normally absent

Table 21.3

'genetic distance' (the higher the value, the more distant the relationship).

The genetic distances of *Apodemus sylvaticus granti* and *Apodemus sylvaticus fridariensis* are given as boxed figures on the map. The upper figure indicates the genetic distance between the island population and the Norwegian mainland species and the lower figure that between the island population and the Scottish mainland species.

a) State ONE quantitative and ONE qualitative difference between the fieldmouse native to Fair Isle and the one from the Scottish mainland. (2)

b) Suggest why scientists do not classify the three subspecies given in table 21.3 as separate species. (1)

c) (i) Berry suggests that the first population of *Apodemus* became established on Yell after being brought from Norway on Viking ships. What evidence from the data supports this theory? (1)

(ii) Fossil evidence shows that *Apodemus* became extinct in Britain during the last Ice Age (about a million years ago). When the ice melted, the Shetland Isles became separated from the mainland and have remained so ever since. Does this information lend support to, or cast doubt on, Berry's theory? Give a reason for your answer. (2)

d) Fair Isle is thought to have been colonised by *Apodemus* from Yell. Explain the genetic distance data referring to *Apodemus sylvaticus fridariensis* in terms of genetic drift. (2)

e) St. Kilda and several Hebridean Islands (e.g. Lewis, Mingulay and Rhum) each possesses its own distinct subspecies of *Apodemus*. Suggest how this could have come about. (2)

(10)

22 Adaptive radiation

Darwin's finches

In 1835, Charles Darwin visited the Galapagos Islands which lie about 600 miles west of the South American mainland. He found them to be inhabited by many different species of finch which varied greatly in **beak size** and **shape** according to diet. On the other hand he found the mainland to have only one species of finch. It is thought therefore that the islands were colonised by a flock of the mainland species carried there by freak weather conditions.

An organism's **ecological niche** is its status within a **community**. Although often applied solely to the organism's mode of feeding, this term refers to its whole way of life including important factors such as habitat, competitors and enemies.

On the mainland, the one species of finch was unable to exploit ecological niches already occupied by other birds (e.g. woodpeckers, warblers etc.). The finches that arrived on the Galapagos found themselves in an environment

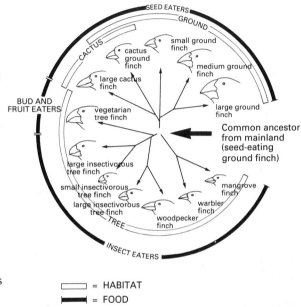

Figure 22.1 *Adaptive radiation in finches*

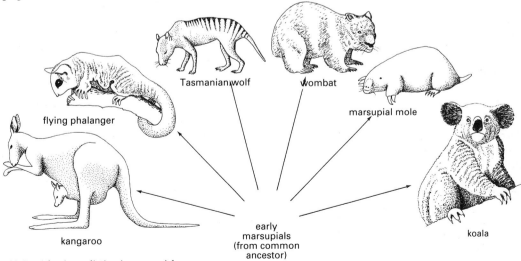

Figure 22.2 *Adaptive radiation in marsupials*

lacking competitors and offering a wide range of unoccupied ecological niches. Reduced selection pressure enabled these finches to increase in number and occupy the islands.

Sub-populations of these finches became *isolated* by barriers (see page 111) and each group underwent its own course of evolution by becoming **diversified** and **adapted** to suit an available ecological niche (see figure 22.1).

This evolution over a long period of time of a group of related organisms along several different lines by adapting to a wide variety of environments is called **adaptive radiation**.

Marsupials

In the absence of competition from placentals, the early marsupial colonists of Australia were able to undergo adaptive radiation. Over a long period of time the variety of types present today evolved (see figure 22.2). Similar ecological niches in other ecosystems are occupied by placentals (see table 22.1).

British buttercups

Table 22.2 refers to three species of British buttercup. Although two or more of these species may be found growing on the same ploughed field or ridged grassland, each is restricted to its own *particular habitat* as shown in figure 22.3.

The fact that each of these closely related species of buttercup has become adapted to suit one particular ecological niche provides evidence that they have undergone adaptive radiation from a common ancestor with each species evolving along a separate line.

Homology and divergent evolution

Structures are said to be **homologous** if they have the same evolutionary origin and are structurally alike. They need not perform the same function.

A group of related organisms that have undergone adaptive radiation are found to possess homologous structures. Mammals (and other terrestrial vertebrates), for example, all

marsupial	ecological niche		example of placental that occupies similar ecological niche in another ecosystem
	habitat	mode of feeding	
kangaroo	grassland	herbivorous (grass)	deer
koala	eucalyptus and gum trees	herbivorous (leaves)	sloth
flying phalanger	trees	omnivorous (fruit and insects)	flying squirrel
marsupial mole	underground burrow	insectivorous (insects and worms)	European mole
Tasmanian wolf	rocky terrain	carnivorous (flesh of birds and mammals	wolf
wombat	underground burrow	herbivorous (grass and roots)	capybara

Table 22.1   *Marsupial and placental occupants of ecological niches*

common name	bulbous buttercup	meadow buttercup	creeping buttercup
scientific name	*Ranunculus bulbosus*	*Ranunculus acris*	*Ranunculus repens*
diploid chromosome number (2n)	16	14	32
symbol in figure 22.3	B	A	R
habitat	dry well-drained soil	damp soil	heavy wet soil

Table 22.2   *Three species of British buttercup*

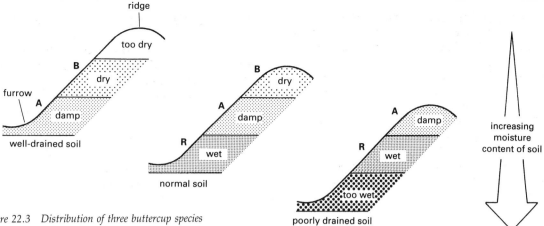

Figure 22.3   Distribution of three buttercup species

possess the **pentadactyl** (five digit) **limb** but in the course of evolution it has become adapted to suit different functions (see figure 22.4). This process is called **divergent evolution** (**divergence**).

Like fossils, the existence of homologous structures provides powerful evidence to support the theory of evolution.

Analogy and convergent evolution

Structures are said to be **analogous** if they perform the same function yet are found to be structurally dissimilar and to have arisen from different evolutionary origins.

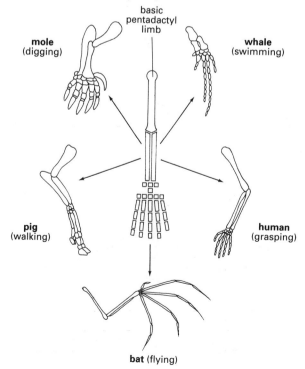

Figure 22.4   Homologous pentadactyl limbs

The wings, eyes and legs of a blackbird, for example, are analogous to those of a locust. Such structures are said to have arisen by **convergent evolution** (**convergence**).

The three small burrowing mammals referred to in table 22.3 all possess powerful digging claws and are similar in appearance. However they are thought to have evolved from separate ancestors within the mammalian group. By becoming adapted to occupy similar ecological niches in their own part of the world, they appear to be much more closely related than they really are. This is a further example of convergent evolution.

common name	reproductive type	native continent
European mole	placental	Europe
strand mole	placental	Africa
marsupial mole	marsupial	Australia

Table 22.3   Convergent evolution of moles

QUESTIONS

---

1 a) (i) Calculate the total volume of each of the human models shown in figure 22.5.

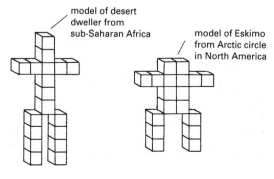

Figure 22.5

123

(ii) Calculate the total surface area of each of the models.

(iii) Relate the difference between the two 'humans' to the environment in which each lives.

b) It is thought that the first humans evolved in a climate much less extreme than either of the two in (a), and that they were intermediate in structure to the above two varieties of humankind.

(i) Explain the meaning of the term adaptive radiation with reference to the above examples.

(ii) Why is it unlikely that humans will evolve into separate species?

2 An octopus eye is very similar to a vertebrate eye. The former however does not possess an inverted retina and develops in a different manner. This indicates evolution from an ancestral line unrelated to the vertebrates.

a) Are these eye types homologous or analogous? Explain your answer.

b) Have they arisen by convergent or divergent evolution?

3 The marsupial wolf of Australia (now nearing extinction) is almost identical in body shape and life style to the placental wolf of the northern hemisphere. However they are not regarded as being close relatives. Explain why.

DATA INTERPRETATION

Figure 22.6 shows a map of the Galapagos Islands. Each figure in brackets refers to the number of different species of finch found on the island.

Table 22.4 gives further information about six of the islands. Table 22.5 and figure 22.7 refer to ground-living finches on two of the islands.

name of island	relative size of island	location of island relative to centre of group	percentage number of finch species only found on this island
Culpepper	small	distant	75
Pinta	small	distant	33
Española	small	distant	67
Isabella	large	central	22.2
Santa Cruz	medium	central	0
Santa Fe	small	central	14.3

Table 22.4

a) Calculate the actual number of finch species which are found only on Culpepper and on no other island. (1)

b) One of the islands has an actual number

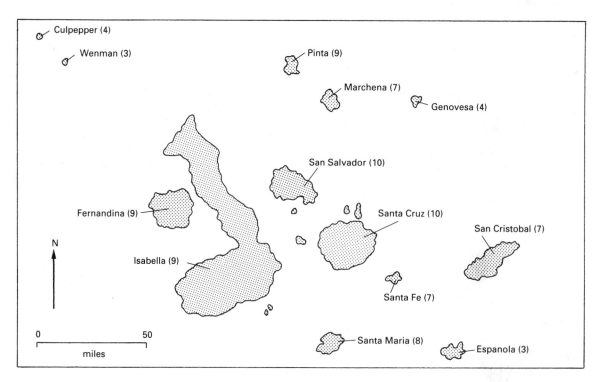

Figure 22.6

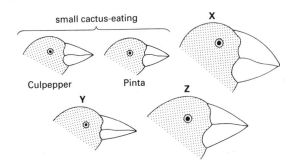

small cactus-eating

Culpepper Pinta

X

Y

Z

Figure 22.7

species of ground finch		island	
body size	food	Culpepper	Pinta
small	cactus	present (11.3 × 9.0)	present (9.7 × 8.5)
large	cactus	absent	present (14.6 × 9.7)
medium	seeds	present (15.0 × 16.5)	absent
large	seeds	absent	present (16.0 × 20.0)

Table 22.5 (numbers in brackets = length × depth of beak in mm)

of two species of finch which are found on it and nowhere else. Identify the island. (1)

c) Which island is populated by finch species which are all found on other islands? (1)

d) (i) In general, what relationship exists between percentage number of finch species found on only one island and that island's geographical location relative to the centre of the group? (1)

(ii) It has been suggested that this relationship could be dependent upon island size. Why is this unlikely? (1)

(iii) Suggest a more likely explanation to account for the relationship. (2)

e) Identify birds X, Y and Z using the information in table 22.5. (1)

f) (i) If a large number of finch type Z were transported from Culpepper to Pinta, which native type on Pinta would face most competition? (1)

(ii) Predict the possible outcome over a long period of time. (1)

(10)

What you should know (CHAPTERS 21–22)

1 The **frequency** of occurrence of a **gene** in a large population mating at random remains **constant** unless the gene pool to which it belongs is affected by **mutation, natural selection, gene migration** or **genetic drift**.

2 The members of a **species** form a natural interbreeding group which is **reproductively isolated** from other species.

3 **Evolution** is a **continuous** process and, as new species appear, other less successful ones become extinct.

4 The process of **speciation** depends on

barriers to gene exchange dividing a population into two or more isolated groups, each of which takes its own course of evolution.

5 **Isolation** has led to **conservation** of some species such as the marsupials.

6 **Adaptive radiation** is the **divergence** over a very long period of time of a group of related organisms along several **different lines** by each becoming adapted to suit a particular ecological niche.

23 Growth differences between plants and animals

Growth

Although growth of a multicellular organism is normally accompanied by an increase in body size, this criterion alone is inadequate to define growth. For example, a middle-aged person who gains several kilograms of fat round the waist and then sheds it by dieting is not showing true growth.

Growth is the **irreversible increase** in the dry mass of an organism and is normally accompanied by an increase in **cell number**. Fresh mass is a less reliable indicator of growth due to temporary fluctuations in an organism's water content.

Meristems

Growth occurs all over a developing animal's body. In plants, growth is restricted to regions called **meristems**. A meristem is a group of undifferentiated plant cells which are capable of dividing repeatedly.

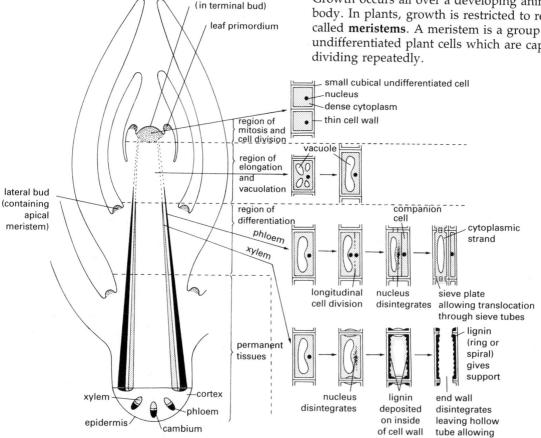

Figure 23.1 Growth of a shoot

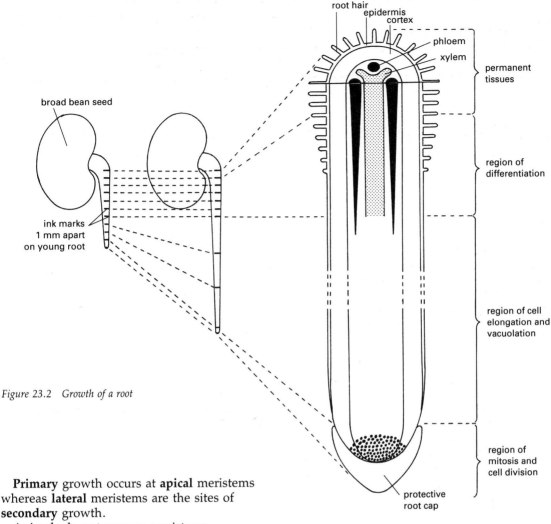

Figure 23.2   *Growth of a root*

Primary growth occurs at **apical** meristems whereas **lateral** meristems are the sites of **secondary** growth.
   Animals do not possess meristems.

Apical meristems

An apical meristem is found at a root and a shoot tip (apex). Increase in length of a root or shoot depends on both the formation of new cells by the apical meristem and elongation of these new cells. Figure 23.1 shows how growth occurs at a shoot apex. Each new cell formed by mitosis and cell division in the meristem becomes **elongated, vacuolated** and finally **differentiated**.
   Differentiation is the process by which an unspecialised cell becomes altered and adapted to perform a special function as part of a permanent tissue. Some cells, for example, become xylem vessels by developing into long hollow tubes supported by lignin. Others become phloem tissue consisting of sieve tubes and companion cells.

Investigating growth in a young root

The experiment illustrated in figure 23.2 shows how the same pattern of events is repeated in a growing root. When the young root is marked with ink at 1 mm intervals and allowed to grow for a few days, the space between the tip and mark '1' is found to remain unaltered since this region consists of the protective root cap and the region of mitosis and cell division.
   However marks 1, 2, 3 and 4 do become spread out since cells in this region are undergoing elongation and vacuolation. Above mark 4 there is no change in the spacing of the ink marks because here the cells are fully elongated and are undergoing differentiation (into xylem, phloem, root hairs etc.).

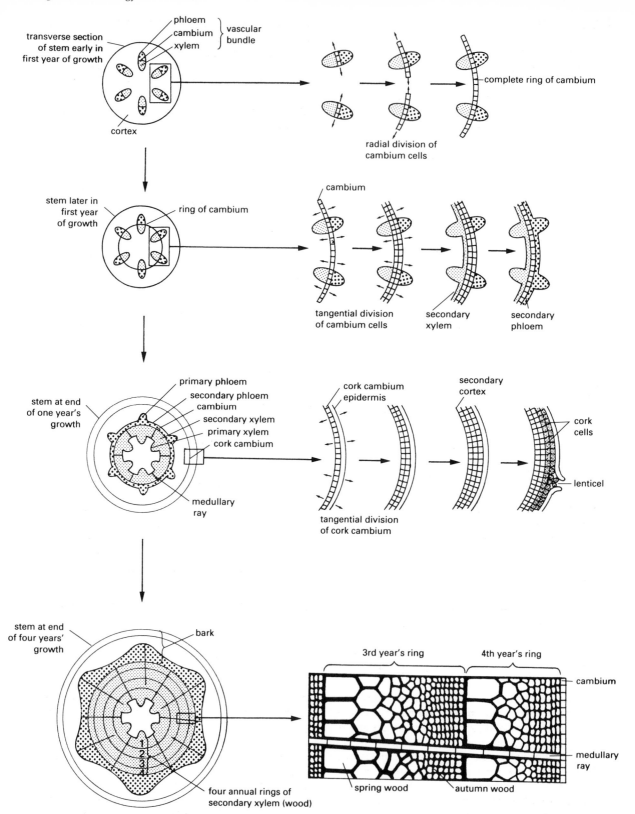

Figure 23.3 Secondary thickening in a woody stem

Lateral meristems

Whereas annual plants die after one year's growth, perennial plants continue to grow year after year. In order to support the new tissues formed at the shoot tips, the plant becomes thicker.

The increase each year in the thickness of the stem (and root) of a woody perennial (e.g. a tree) is produced by secondary growth (**secondary thickening**) controlled by two lateral meristems, the **cambium** and the **cork cambium**.

Cambium

Figure 23.3 shows the process of secondary thickening in a stem. A complete ring of **secondary xylem** (and a less obvious ring of secondary phloem) arise annually from the meristematic ring of cambium: xylem to the inside, phloem to the outside. (Such differentiation is under genetic control, see chapter 25).

The **medullary rays**, which appear first amongst the secondary xylem and later amongst the secondary phloem, each consist of a row of relatively unspecialised cells (called **parenchyma**) which arise from the cambium. Their function is lateral transport of water and mineral salts from the xylem to the other living cells.

Cork cambium and lenticels

As the internal tissues increase in thickness, the outer layers become stretched. However a ring of cork cambium forms outside the cortex. This makes new **cortex** cells to its inside and dead impervious **cork** cells to its outside, thus preventing rupture of the outer tissues.

Lenticels are tiny pores in the epidermis where the cork cells are loosely packed. They allow gaseous exchange to occur between the atmosphere and the inner living cells.

Annual rings and grain of timber

In temperate regions, cambium is most active in spring and large xylem vessels (**spring wood**) are formed. These wide vessels are able to transport the large quantities of water and nutrients needed in spring for leaf growth and expansion.

By the time autumn arrives, the cambium is less active and produces smaller xylem vessels (**autumn wood**). Cambium is inactive in winter.

This growth pattern is repeated annually allowing each year's ring of woody xylem to be clearly distinguished and the age of the tree to be easily calculated.

The thickness of an annual ring is dependent upon the growth conditions that prevailed during that year. A narrower ring is formed during cold and/or dry conditions or infestation of leaves by insects. A wider ring indicates a year of warm and/or wet weather.

Grain of timber

When timber is cut longitudinally through the vascular tissue, the rings show up as patterns called the **grain** of the wood (figure 23.4).

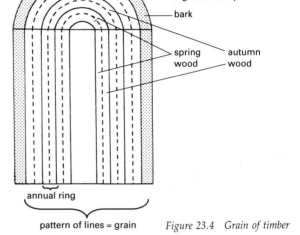

Figure 23.4   Grain of timber

In older trees the innermost xylem vessels (wood) lose their water-conducting capacity and become filled with gums and tannins. This central wood is called **heartwood** to distinguish it from the outer water-conducting **sapwood**.

Regeneration in angiosperms and mammals

Regeneration is the process by which an organism *replaces* lost or damaged parts. Flowering plants (**angiosperms**) have extensive powers of regeneration. A piece of stem or root cut from many types of plant will develop roots (and buds) and eventually regenerate the entire plant.

Horticulturalists employ artificial methods of vegetative propagation (asexual reproduction) such as taking cuttings to increase their supply of a desirable type of plant. The success of this process rests solely on the ability of many types of angiosperm cuttings to regenerate their

23.6), mending of broken bones, replacement of blood after loss and regeneration of damaged liver.

So great is the liver's potential for regeneration that half of a liver can regain its full size within three months. This is made possible by the fact that liver cells are less highly differentiated than most other body cells and therefore the genes which control the synthesis of the essential growth factor can become 'switched on' when required. Cells affected in this way divide and behave like young cells until the liver has reached its full size. The genes then become 'switched off' again. (The switching on and off of genes is discussed more fully in chapters 25 and 26.)

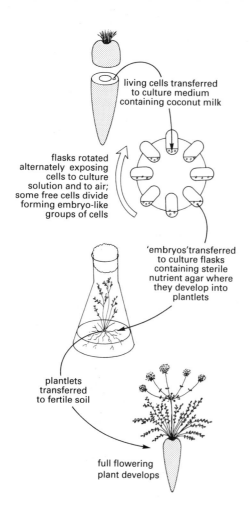

Figure 23.5 Tissue culturing

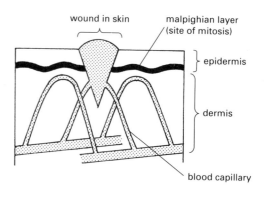

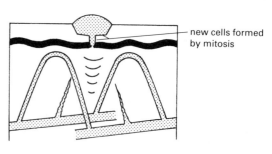

missing parts. Such regeneration is often promoted by the application of rooting powders containing growth substances (see page 159) to the cut surface.

Tissue cultures

Thousands of identical plants can be produced by growing **tissue cultures**. Early work in this field involved the use of carrot plants (see figure 23.5). This technique now enables commercial growers to mass produce clones of plants such as pineapple, rose, orchid and oil palm.

Mammals

Mammals have only limited regenerative powers. These are restricted to healing of wounds (figure

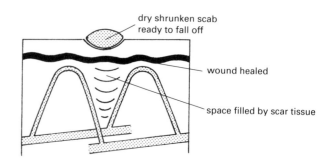

Figure 23.6 Wound healing

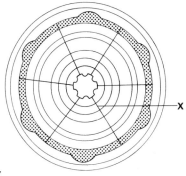

Figure 23.7

QUESTIONS

1 Define:
   a) growth,
   b) regeneration.
2 a) What is a meristem?
   b) Identify TWO regions of a plant which possess an apical meristem.
   c) State the function of meristematic cells.
3 a) Which TWO processes produce downward growth of a root?
   b) In an experiment, a young root was inked at 1 mm intervals as shown in figure 23.2. After 4 days of further growth, the spaces between the marks were measured and recorded as shown in table 23.1.

space between:	distance (mm)
root tip and mark 1	1
marks 1 and 2	6
2 and 3	7
3 and 4	6.5
4 and 5	1
5 and 6	1
6 and 7	1
7 and 8	1
8 and 9	1
9 and 10	1

Table 23.1

   (i) Present the data as a bar graph.
   (ii) Draw TWO conclusions from the data.
4 State THREE differences between a xylem vessel and a meristematic cell.
5 What results from radial division of the original cambium cells in a stem's vascular bundles?
6 The following questions refer to figure 23.7 of a transverse section of a stem.

a) What age is the specimen?
b) Name **X** and state its function.
c) During which year were the conditions most suitable for the development of wood?
d) Identify the year during which the tree's leaves may have been infested with insect larvae.
Explain how you arrived at your answer.
7 Figure 23.8 shows a log cut transversely and longitudinally.

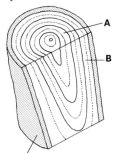

Figure 23.8   log of timber cut in two planes

a) From which tissue were **A** and **B** formed?
b) State ONE difference between the 'cells' at **A** and **B**.
c) What name is given to wood patterns such as those shown in this log's longitudinal plane?
8 Suggest why trees growing near the equator do not contain distinct annual rings.

24 Growth patterns

Growth curve

Measuring the dry mass of an organism involves removing all of its water. This often proves impractical and therefore growth is usually investigated by measuring a variable factor such as fresh weight or height over a period of time. When graphed the results give a **growth curve**. This commonly takes the form of a **sigmoid** (S-shaped) curve as shown in figure 24.1.

The exact form that this basic growth curve takes depends on the organism involved.

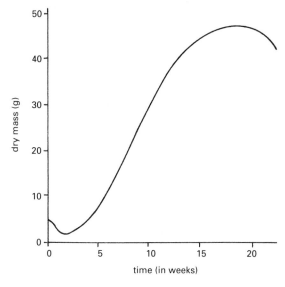

Figure 24.2   Annual plant growth curve

Annual plant

Figure 24.2 shows the growth curve for a pea plant. At first there is a decrease in mass as food reserves are used during seed germination. Soon photosynthesis begins and growth proceeds in the normal manner.

Annual plants only grow for a short time (less than a year) and then stop. Towards the end of the growing season the curve flattens out and then quickly goes into decline when the seeds are dispersed.

Tree

Figure 24.3 shows the growth curve for a tree which is a perennial plant and continues to grow for many years.

Each year, spring and summer provide the best

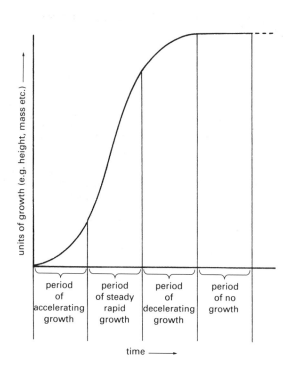

Figure 24.1   Sigmoid growth curve

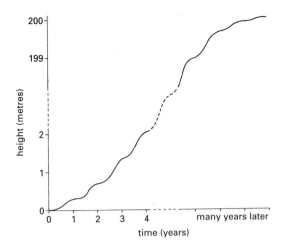

Figure 24.3   *Perennial plant growth curve*

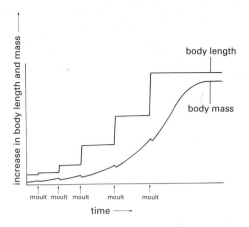

Figure 24.5   *Insect growth 'curves'*

growing conditions. During winter, growth slows down and often comes to a complete halt.

The growth pattern is therefore found to take the form of a series of annual sigmoid curves whose overall shape is, in turn, sigmoid.

Human

Figure 24.4 shows the growth curve for a human. Although basically sigmoid, it has two phases of rapid growth (called **growth spurts**) instead of one.

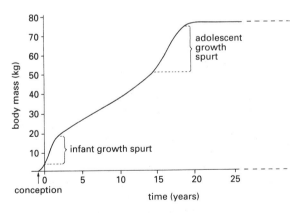

Figure 24.4   *Human growth curve*

Insect

Figure 24.5 shows the growth pattern of an insect (e.g. locust). Insects possess a hard inelastic outer skin called an **exoskeleton** which prevents continuous increase in body size. This skin must be periodically shed to allow growth to occur.

A new skin forms beneath the old one and when the latter is cast off during **moulting (ecdysis)**, the insect quickly inflates its body before the new skin hardens. This produces an abrupt increase in body length giving the graph its step-like appearance.

Such discontinuous growth only applies to measurements of body size (e.g. length). The mass of the insect increases steadily between moults and, if graphed, gives a curve with a fairly sigmoid appearance.

Comparison of plant and animal growth

Table 24.1 summarises the growth differences between advanced plants and animals.

plant (e.g. angiosperm)	animal (e.g. mammal)
growth only occurs at meristems	growth occurs all over body
growth (increase in size) continues throughout life	growth (increase in size) stops on reaching adulthood
regenerative powers are very extensive	regenerative powers are very limited

Table 24.1   *Comparison of plant and animal growth*

QUESTIONS

1 The data in tables 24.2 and 24.3 refer to two different types of living organism.

time after hatching or germinating (days)	length (mm)
0.0	80
6.0	80
6.1	150
11.0	150
11.1	200
15.0	200
15.1	260
20.0	260
20.1	500
30.0	500
30.1	760
40.0	760

Table 24.2

time after hatching or germinating (weeks)	dry mass (g)
0	6
2	2
4	6
6	12
8	20
10	30
12	38
14	44
16	48
18	50
20	50
22	49
24	42

Table 24.3

a) Using two sheets of graph paper, plot a separate line graph of each set of data (with time on the x axis in both cases).

b) Identify which graph depicts the growth curve of (i) an annual plant, (ii) an insect.
c) Give TWO reasons for your choice of answer to (b).
2 State THREE ways in which the growth of a rose bush differs from that of a rabbit.

What you should know (CHAPTERS 23–24)

1 Growth is the **irreversible** increase in the **dry mass** of an organism.
2 In plants, growth is restricted to regions called **meristems**.
3 **Primary** plant growth occurs at **apical meristems** (root and shoot tips) where newly formed cells become **elongated, vacuolated** and **differentiated**.
4 **Secondary** growth occurs in perennial plants at lateral meristems (**cambium** and **cork cambium**). Cambial activity produces **annual rings** of secondary xylem and phloem.
5 Animals do not possess meristems. Instead growth occurs **all over** a developing animal's body.
6 Flowering plants (angiosperms) have extensive powers of **regeneration** whereas mammals have only limited powers.
7 Investigations into growth often involve measuring a **variable factor** such as fresh weight, height, length etc.
8 A graph of the results normally takes the form of a **sigmoid (S-shaped)** growth curve.
9 Growth patterns vary from one type of organism to another.

25 Genetic control

Role of genes

The master plan for the development of an organism is contained in the nuclei of its cells. To function, every living cell must possess a variety of **proteins**, some to form its basic structure, others to act as enzymes which control metabolic processes.

These essential proteins are synthesised according to the base sequence encoded in the cell's DNA (see page 40). A particular segment of DNA called a **gene** codes for each protein (or polypeptide). Thus the structure and function of a cell is determined and controlled by its genes.

Differentiation

Since all the cells in a multicellular organism have arisen from the original zygote by repeated mitotic division, every cell possesses all the genes for constructing the whole organism. However as development proceeds, different cells become **specialised** to perform different functions (see page 127).

This is made possible by the fact that within a particular type of differentiated cell, only certain genes continue to operate whilst in other types of differentiated cells different groups of genes remain functional.

Reversibility of differentiation

In some cases, when the nucleus is taken out of a fully differentiated cell (e.g. intestine) of an amphibian tadpole and transplanted into a fertilised egg (from which the nucleus has been removed), the egg will still grow into a fertile adult.

This provides evidence that a differentiated cell does contain all of the organism's genes and that genes which had been repressed ('switched off') were able to be switched on again.

Two categories of genes

The genes present in all cells can be classified into two distinct categories.

The first group are expressed ('switched on') in all cells and code for **vital metabolites** necessary for the biochemical processes basic to life (e.g. **enzymes** which control respiratory pathways).

The second group of genes are only switched on in a particular type of cell where they code for proteins *characteristic* of that cell type and no other (e.g. **insulin** made by pancreas cells). Many of these proteins also bring about **specific modifications** of cell structure and determine its characteristic features. The other genes not needed by the cell remain permanently switched off.

Genetic control of cells

Blood

Blood contains red blood cells and several types of white blood cell. The particular features possessed by a fully **differentiated** blood cell depend upon which of its genes are switched on and which were repressed during its differentiation from an undifferentiated ('stem') cell as shown in figure 25.1

Parenchyma

Parenchyma are large, thin-walled, vacuolated cells which make up the 'ground' tissue of all plant organs. They contain organelles called **plastids** (which are formed from proplastids) and vary according to their pigmentation.

Various types of parenchyma cells exist. The particular features that each possesses depend upon which of its genes are switched on and

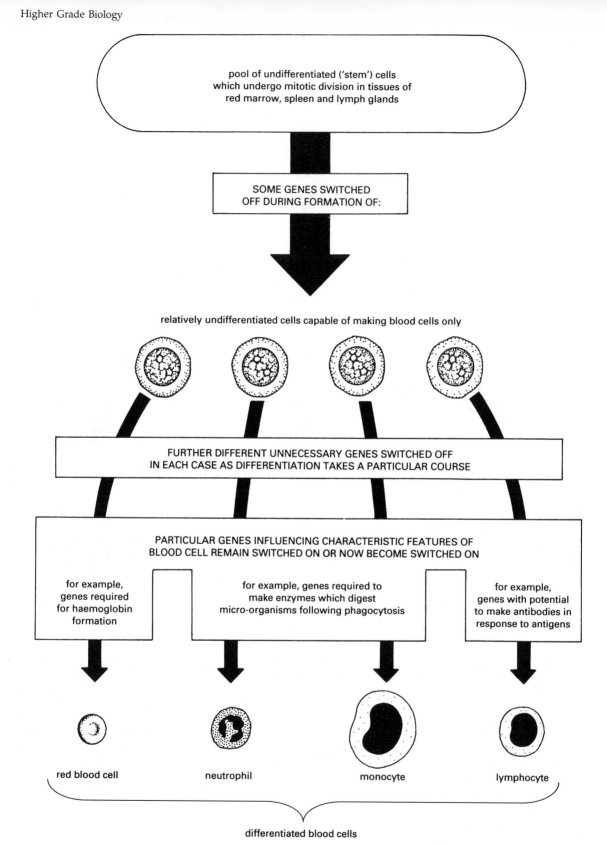

Figure 25.1 Genetic control of blood cells

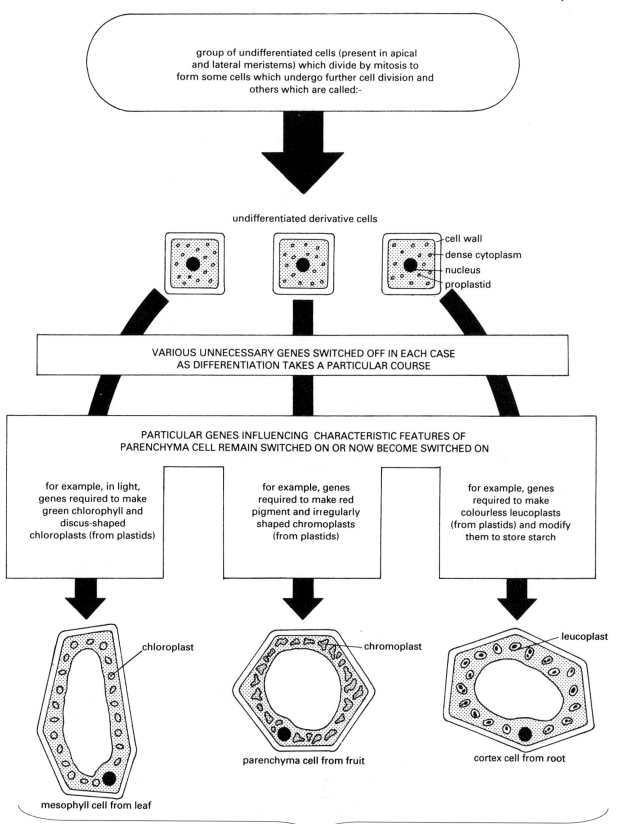

group of undifferentiated cells (present in apical and lateral meristems) which divide by mitosis to form some cells which undergo further cell division and others which are called:-

undifferentiated derivative cells

cell wall
dense cytoplasm
nucleus
proplastid

VARIOUS UNNECESSARY GENES SWITCHED OFF IN EACH CASE AS DIFFERENTIATION TAKES A PARTICULAR COURSE

PARTICULAR GENES INFLUENCING CHARACTERISTIC FEATURES OF PARENCHYMA CELL REMAIN SWITCHED ON OR NOW BECOME SWITCHED ON

for example, in light, genes required to make green chlorophyll and discus-shaped chloroplasts (from plastids)

for example, genes required to make red pigment and irregularly shaped chromoplasts (from plastids)

for example, genes required to make colourless leucoplasts (from plastids) and modify them to store starch

chloroplast

chromoplast

leucoplast

mesophyll cell from leaf

parenchyma cell from fruit

cortex cell from root

Figure 25.2   Genetic control of parenchyma

differentiated parenchyma cells

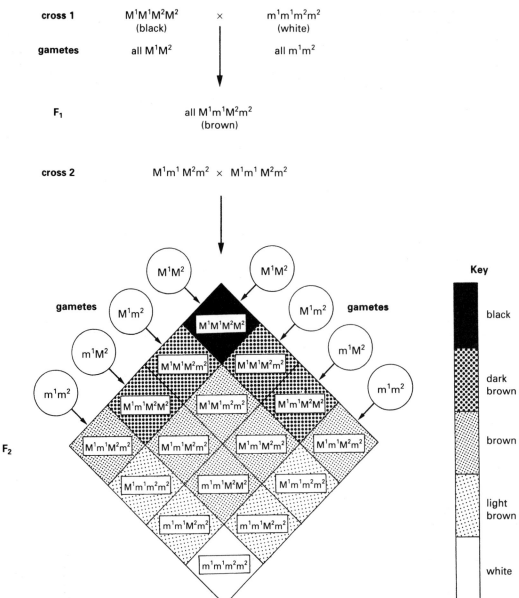

Figure 25.3 *Genetic control of skin colour*

which were repressed during its differentiation from a meristematic cell. This is shown in figure 25.2.

Compared with other differentiated plant cells such as xylem vessels, parenchyma cells are *relatively unspecialised* and many are found to be potentially meristematic.

If cultured in suitable growth conditions, a parenchyma cell can often divide to form a group of cells called a callus which will develop into a complete plant. This shows that genes which were repressed in the original parenchyma cell can become switched on again.

Genetic control of skin colour

People vary in skin colour from very dark to very light depending upon the amount of the pigment **melanin** that is present in their skin cells.

Ignoring the temporary suntanning effect of exposure to ultraviolet radiation, the amount of melanin deposited in skin cells is a genetically controlled characteristic.

Such variation in skin colour can be explained on the basis that it is determined by two independent genes each with two alleles as shown in table 25.1.

	dominant allele	recessive allele
gene 1	M^1 (codes for melanin)	m^1 (does not code for melanin)
gene 2	M^2 (codes for melanin)	m^2 (does not code for melanin)

Table 25.1   Genes for skin colour

If these two genes work in such a way that their effects are *additive* then each dominant allele would code for a quantity of melanin equal to 25 per cent of the maximum possible amount that can occur in human skin.

Thus a person with the genotype $M^1M^1M^2M^2$ would possess the maximum amount of pigment and be 100 percent black, person $M^1M^1M^2m^2$ would be 75 per cent black (i.e. dark brown) and so on to person $m^1m^1m^2m^2$ who would be white.

It is certainly true that a cross between a black person of pure West African descent and a white person of pure North European descent results in F_1 children possessing intermediate brown skin. In addition, crosses between such brown F_1 individuals produce offspring displaying a range of skin colours. This is summarised in figure 25.3.

However, since there are many more than five tones of human skin pigmentation, figure 25.3 only represents a simplified version of the true picture. Studies of many crosses suggest that in reality skin colour is determined by four (or more) genes whose dominant alleles have an additive effect.

This type of inheritance pattern where one characteristic is determined by several different genes is called **polygenic**.

Genetic control of eye colour

The colour of the iris of the human eye is often said to be controlled by a single gene where the brown allele (B) is dominant to the recessive blue allele (b). However the fact that it is possible for two blue-eyed parents to produce a brown-eyed child shows that this explanation is over-simplified.

The colour of an iris is determined by the amount of melanin that it possesses. Melanin is a light-absorbing pigment. An iris rich in melanin is dark in appearance (e.g. dark brown) since it absorbs much light. An iris with little or no melanin is light in appearance (e.g. pale blue) since it reflects light.

Iris colour is thought to be a further example of polygenic inheritance. It can be explained on the basis that it is controlled by three genes

genotype(s)	phenotype	number of alleles adding melanin
$B^1B^1B^2B^2B^3B^3$	deep brown	6
$B^1B^1B^2B^2B^3b^3$, $B^1B^1B^2b^2B^3B^3$ etc.	dark brown	5
$B^1B^1B^2B^2b^3b^3$, $B^1B^1B^2b^2B^3b^3$ etc.	medium brown	4
$B^1b^1B^2b^2B^3b^3$, $B^1B^1b^2b^2b^3b^3$ etc.	light brown	3
$B^1b^1B^2b^2b^3b^3$, $B^1b^1b^2b^2B^3b^3$ etc.	deep blue	2
$B^1b^1b^2b^2b^3b^3$, $b^1b^1B^2b^2b^3b^3$ etc.	medium blue	1
$b^1b^1b^2b^2b^3b^3$	pale blue	0

Table 25.2   Genetic control of eye colour

whose dominant alleles code for melanin and have an **additive** effect as shown in table 25.2.

Thus there is a 1 in 8 chance of a brown-eyed child being produced by the two blue-eyed parents shown in figure 25.4.

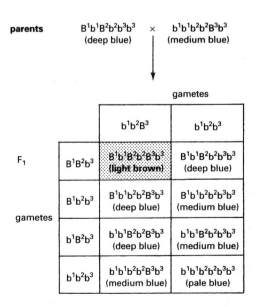

Figure 25.4   Genetic control of eye colour

Genetic control of leaf shape

Each leaf begins life as a small group of cells called a leaf **primordium** produced by an apical meristem (see page 126). The blade of the leaf develops from **meristematic** cells at the edges of the leaf primordium (see figure 25.5).

Equal activity of these cells during growth and development produces a leaf with a regular shape (e.g. privet). Unequal activity produces a leaf with an irregular shape (e.g. oak).

The activity of these meristematic cells in a developing leaf is controlled by growth substances which are in turn controlled by the plant's genes. Thus the basic shape of a leaf is genetically determined and is a characteristic feature of each species of plant.

QUESTIONS

---

1 When some plant cells are exposed to light their leucoplasts may develop into chloroplasts containing green chlorophyll.
a) Suggest what happens within the cells to bring about this change.
b) Why is it of advantage to the plant that plastids in non-illuminated cells do not become chloroplasts?

2 For the sake of simplicity, assume that human skin colour is determined by the two genes referred to in figure 25.3. Show in diagrammatic form the outcome of a cross between $M^1m^1M^2m^2$ and $m^1m^1m^2m^2$. Clearly state the genotype and phenotype of each type of offspring produced.

3 Assume that the colour of the iris of the human eye is determined by three genes as explained on page 139.

What is the chance of each of the following sets of blue-eyed parents producing a brown-eyed child? Construct a Punnett square for each cross to show how you arrived at your answers:
a) $b^1b^1B^2b^2b^3b^3 \times B^1b^1b^2b^2B^3b^3$
b) $b^1b^1B^2b^2b^3b^3 \times B^1b^1b^2b^2b^3b^3$
c) $B^1b^1B^2b^2b^3b^3 \times B^1b^1b^2b^2B^3b^3$

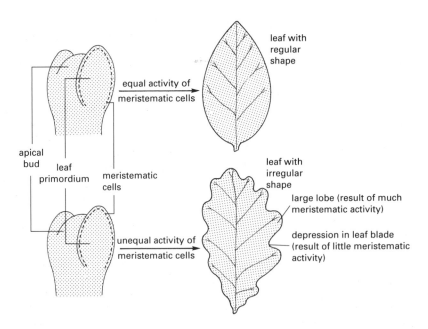

Figure 25.5 Genetic control of leaf shape

26 Control of gene action

Some proteins are only required by a cell under certain circumstances. To prevent resources being wasted, the genes that code for these proteins are switched on and off as required.

Enzyme induction

Lactose is a type of sugar which is digested to glucose and galactose by the enzyme β-**galactosidase**. The bacterium, *Escherichia coli*, is found to produce β-galactosidase when lactose is present in its nutrient medium but fails to do so when lactose is absent. Somehow the presence of lactose switches on the gene which codes for β-galactosidase.

This process of switching on a gene only when the enzyme that it codes for is needed is called **enzyme induction**.

Operon hypothesis of gene action

An **operon** consists of one or more **structural** genes (containing the DNA code for the enzyme in question) and a neighbouring **operator** gene which activates the structural gene(s). The operator gene is, in turn, affected by a repressor molecule coded for by a **regulator** gene situated further along the DNA chain.

On this basis, enzyme induction can be explained as shown in figure 26.1. During the induction of β-galactosidase, lactose (the **inducer**) prevents the repressor molecule from inactivating the operator gene. The structural gene is therefore switched on and production of β-galactosidase proceeds.

When all the lactose has been digested, the repressor molecule becomes free again and switches the gene off, thereby preventing resources such as amino acids and energy being used up needlessly.

This hypothesis, first put forward by two scientists called Jacob and Monod, is now supported by extensive experimental evidence from work done using bacteria. However the mechanism by which genes are switched on and off in higher organisms such as humans is not yet fully understood.

Genetic control of enzyme production

Each stage in a metabolic pathway is controlled by an **enzyme**. The production of each enzyme is in turn determined by a particular gene as shown in the imaginary example in figure 26.2.

If one of these genes undergoes a **mutation**, then it may be no longer able to code the correct message for the production of its particular enzyme. In the absence of this enzyme, the pathway becomes disrupted.

Normally the mutated allele of a gene is recessive to the normal allele and only shows its effect in the phenotype of a person who has inherited the mutated allele from both parents. Such a homozygous person suffers a disorder known as an **inborn error of metabolism**.

Phenylketonuria (PKU)

Phenylalanine and tyrosine are two amino acids that humans obtain from protein in their diet. During normal metabolism, excess phenylalanine undergoes the pathway shown in figure 26.3. **Phenylketonuria (PKU)** is a hereditary disorder caused by a genetic defect which disrupts this metabolic pathway. An affected person lacks the normal allele of the gene required to make enzyme 1 (phenylalanine hydroxylase). Phenylalanine is therefore no longer converted to tyrosine. Instead it undergoes alternative metabolic pathways producing toxins which

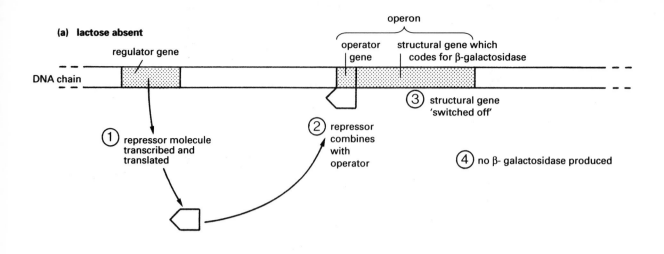

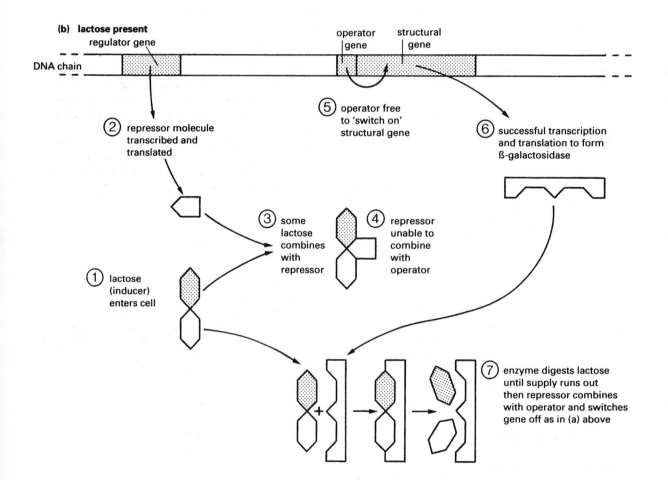

Figure 26.1   Operon hypothesis of gene action

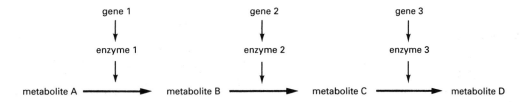

Figure 26.2 Enzyme-controlled metabolic pathway

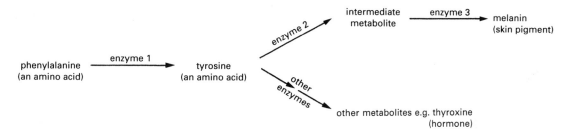

Figure 26.3 Normal fate of phenylalanine

affect the metabolism of brain cells and severely limit mental development.

One of these chemicals (phenylpyruvic acid) is excreted in urine allowing the disorder to be diagnosed. Although a sufferer of PKU has lighter skin pigmentation than normal, he or she is not an albino since some tyrosine is present in the diet for enzyme 2 to act upon.

PKU is a fairly common disorder. It occurs in about 1 in 10 000 of the UK population. However thanks to widespread screening of newborn babies, followed by a lifelong phenylalanine restricted diet for sufferers, its worst effects have now been reduced to a minimum. It is possible that PKU may be cured in the future using genetically engineered cell implants.

Albinism

Owing to a genetic defect, albinos are unable to make enzyme 3 in the pathway shown in figure 26.3 and therefore fail to synthesise melanin.

As a result an albino's skin is pink and fails to tan, the iris is pale blue or pink and the hair is pure white. However albinos suffer no medical problems provided that they avoid exposure to ultraviolet radiation.

Human development at puberty

Hormones (chemical messengers) are produced by an animal's **endocrine** (ductless) **glands** and secreted into the bloodstream. When a hormone reaches a certain (target) tissue, it brings about a specific effect.

The **pituitary gland** (see figure 26.4) is an endocrine gland which produces many hormones. Two of these, **follicle stimulating hormone (FSH)** and **luteinising hormone (LH)**, are called the **gonadotrophic** hormones because their target organs are the **gonads** (reproductive organs).

In certain pituitary cells, the appropriate genes are switched on and code the information needed for the production of FSH and LH (both of which are proteins). These hormones are stored in cell vesicles awaiting the signal for release.

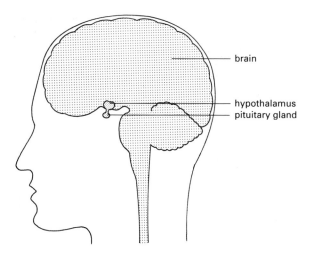

Figure 26.4 Pituitary gland

143

	female	male
average age at puberty	10–14	13–16
reproductive organ (gonad) stimulated by gonadotrophic hormones	ovary	testis
effects of stimulation	onset of menstrual cycle and ovulation (egg release), release of sex hormones (oestrogen and progesterone) by ovary	production of sperm, release of sex hormone (testosterone) by testis
effects of sex hormones on body tissues	increase in height and weight (growth spurt), development of uterus and vagina, development of breasts, growth of pubic hair	increase in height and weight (growth spurt), development of penis and scrotum, deepening of voice, growth of facial and pubic hair

Table 26.1   Puberty

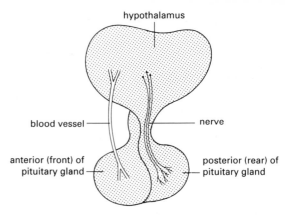

Figure 26.5   Hypothalamus

Puberty is the age at which the reproductive organs reach maturity and a child becomes an adult. It is at this time that the pituitary gland begins to release the gonadotrophic hormones which then bring about the effects described in table 26.1.

Genetic control of hormone production

The **hypothalamus** (see figure 26.5) is a region of the brain connected to the anterior part of the pituitary gland by tiny blood vessels.

The hypothalamus contains secretory cells which make **releaser** hormones. These are transported in the bloodstream to the pituitary,

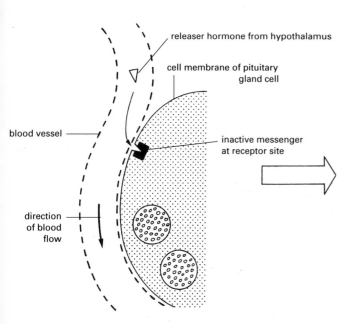

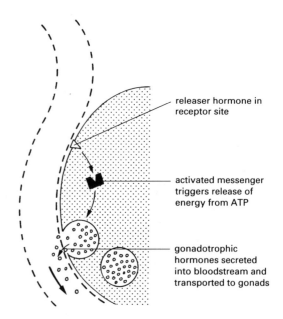

Figure 26.6   Action of releaser hormones on pituitary cell

telling it which hormones to secrete.

The release of both pituitary gonadotrophic hormones is controlled by a single releaser hormone from the hypothalamus. This hormone is a peptide consisting of only ten amino acids. At puberty the gene which codes for this releaser hormone becomes switched on in certain secretory cells in the hypothalamus.

Once synthesised, the hormone is transported to the anterior pituitary gland where it becomes attached to **receptor sites** on the outside of cells containing FSH and LH. It is thought that inactive messengers inside the cells now become active and promote the release of FSH and LH as shown in figure 26.6.

Steroid hormones

Hormones made of protein exert their effect from outside the cell of a target tissue. In contrast, **steroid** hormones (fat soluble lipids) can pass through the cell membrane and influence the target cell's genes.

The **sex** hormones **oestrogen, progesterone** and **testosterone** (see table 26.1) are steroid hormones. They bring about the changes that occur in body tissues during puberty (e.g. growth of pubic hair) by activating certain genes in the cells of the appropriate tissues.

Once switched on, the gene transcribes the mRNA needed to produce a specific protein (e.g. keratin for hair). A simplified version of this process of genetic control is shown in figure 26.7. It is thought that the switching on mechanism may work in a similar way to the operon system described on page 141.

Genetic control of flowering

Many species of plants stop producing leaf buds and start making floral buds at a certain time of the year.

This change is brought about by a hormone which causes those genes controlling vegetative features to become repressed and those controlling floral features (e.g. ovaries, stamens, petals etc.) to become switched on.

In many plants, production of the hormone is dependent upon the plant being exposed to the appropriate **photoperiod** (see chapter 30).

Two-way interaction

The fact that genes can be switched on and off in response to external factors such as hormones and length of photoperiod shows that control of gene action is a **two-way interaction** with the nucleus receiving and responding to feedback from the cytoplasm (as shown in figure 26.8).

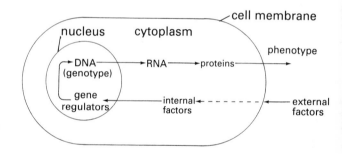

Figure 26.8   Two-way interaction

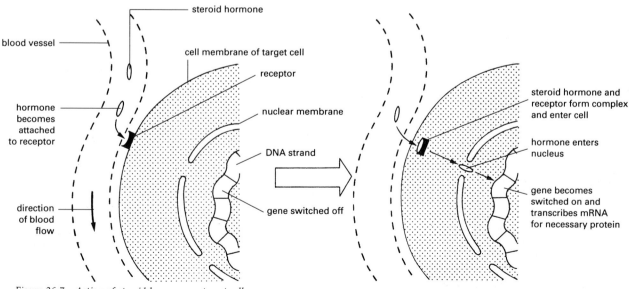

Figure 26.7   Action of steroid hormone on target cell

Thus the developing phenotype is the product of the effect of both internal and external factors acting on the genotype.

QUESTIONS

---

1 a) The following statements refer to events described by the Jacob-Monod hypothesis of gene action. Rewrite them so that each contains the correct word.

(i) The regulator gene produces the $\left\{ \begin{array}{l} \text{repressor} \\ \text{inducer} \end{array} \right\}$ molecule.

(ii) The inducer molecule combines with the $\left\{ \begin{array}{l} \text{operator} \\ \text{repressor} \end{array} \right\}$.

(iii) When the operator is free, the structural gene is switched $\left\{ \begin{array}{l} \text{on} \\ \text{off} \end{array} \right\}$.

b) (i) Explain why *Escherichia coli* only produces the enzyme β-galactosidase when lactose is present in its food.

(ii) What is the benefit of this on/off mechanism to the bacterium?

c) Predict what will happen when lactose enters a cell of *E. coli* which has suffered a nucleotide pair deletion in the centre of the structural gene.

2 Figure 26.9 shows part of a metabolic pathway that occurs in humans. Each stage is controlled by an enzyme. Some of these stages have been given a letter.

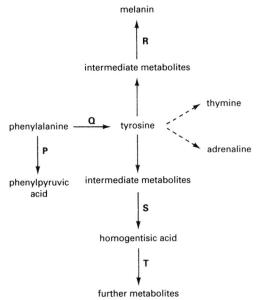

Figure 26.9

a) Explain how a mutation can lead to a blockage in such a pathway.

b) Identify the letter that represents the point of blockage that leads to each of the following disorders:

(i) phenylketonuria,
(ii) albinism,
(iii) alcaptonuria (characterised by accumulation of homogentisic acid which is excreted in urine and turns black in light).

3 The graph in figure 26.10 shows the effect of a phenylalanine meal on a normal person and on a sufferer of phenylketonuria (PKU).

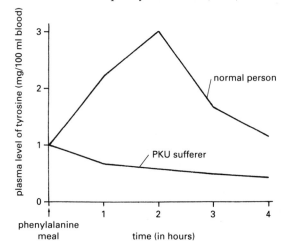

Figure 26.10

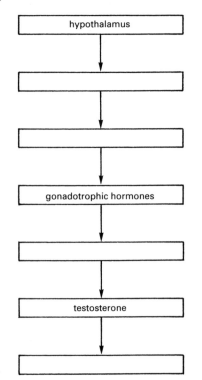

Figure 26.11

a) Explain the initial rise in level of tyrosine in the normal person.

b) Why does the PKU sufferer not show a similar increase?

c) Why does the level of tyrosine in the normal person fall after two hours?

4 a) Copy and complete the flow diagram in figure 26.11 to show the series of events involving hormones which occurs at puberty in a human male.

b) A similar series of events occurs in the human female.

(i) Name ONE gonadotrophic and ONE ovarian hormone involved.

(ii) Which of these is a steroid hormone?

(iii) Describe how a steroid hormone is thought to exert its effect on its target tissue (e.g. skin in armpit about to produce hair).

What you should know (CHAPTERS 25–26)

1 The characteristic features of a cell are controlled by its **genes**.

2 Those genes required to code for vital metabolites are **switched on** in all cells.

3 Only those genes that code for proteins **characteristic of a particular cell type** are switched on in that type of cell. Other genes not needed by the cell remain switched off.

4 To prevent resources being wasted, some genes can be **switched on and off** as required.

5 The **operon** (Jacob-Monod) **hypothesis** states that a structural gene remains switched off while its **operator** gene is inactivated by a **repressor** from a **regulator** gene. The **structural** gene becomes switched on and codes for its protein when an **inducer** prevents the repressor inactivating the operator gene.

6 Production of each **enzyme** is controlled by a particular gene (or group of genes).

7 Each stage in a **metabolic pathway** is controlled by an enzyme.

8 A **mutated** gene is unable to code the information needed to produce its enzyme. Lack of this enzyme may lead to an **inborn error** of metabolism such as PKU.

9 **Hormone** production is also controlled by genes.

10 In humans at **puberty**, the gene for a **releaser hormone** made by **hypothalamus** cells becomes switched on. This hormone promotes the release of **gonadotrophic hormones** by cells in the pituitary gland.

11 Gonadotrophic hormones stimulate gonads to make **sex hormones** which activate genes in body tissues.

27 Hormonal influences on growth – part 1

Animal hormones

In animals, endocrine glands produce **hormones** (chemical messengers) and secrete them directly into the bloodstream. Two of the human body's endocrine glands are shown in figure 27.1.

Role of pituitary

The **pituitary gland** produces a variety of hormones. Two of these play a role in the control of growth and development.

Somatotrophin

The anterior (front) lobe of the pituitary produces human **growth hormone (somatotrophin)**. This

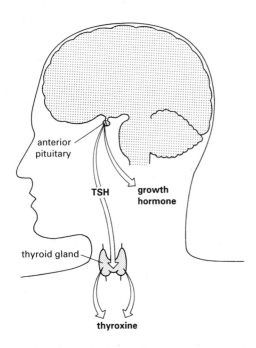

Figure 27.1   *Secretion of hormones affecting growth*

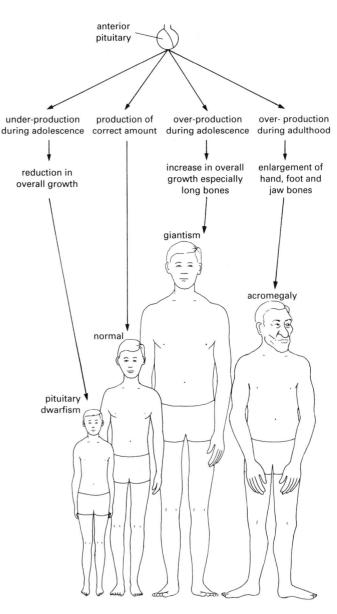

Figure 27.2   *Effects of over- and under-production of somatotrophin*

148

hormone promotes growth by accelerating amino acid transport into the cells of both soft tissues and bones. This permits rapid synthesis of tissue proteins to occur. In particular, it promotes increase in length of long bones during the growing years.

Figure 27.2 shows the effect of over-production and under-production of human growth hormone during adolescence, and the effect of over-production during adulthood.

Thyroid-stimulating hormone

The anterior pituitary also produces **thyroid-stimulating hormone (TSH)** which controls growth and activity of the thyroid gland. This latter endocrine gland responds to TSH by producing **thyroxine** (and other thyroid hormones) which control the body's metabolic processes and therefore affect growth.

If the thyroid gland is inadequately stimulated by TSH then insufficient thyroid hormones are secreted and metabolic rate drops. In extreme cases in children this results in a drastic form of dwarfism called **cretinism** which is characterised by lack of skeletal, sexual and mental growth and development.

Plant growth substances

Plant **growth substances** (hormones) are chemicals produced by plants which speed up, inhibit or otherwise affect growth and development.

Auxins

Early evidence for the existence of plant growth substances came from work done on the coleoptiles of oat seedlings (see figure 27.3). From the series of experiments shown in figure 27.4 the following conclusions were drawn:

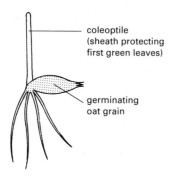

Figure 27.3 Oat seedling

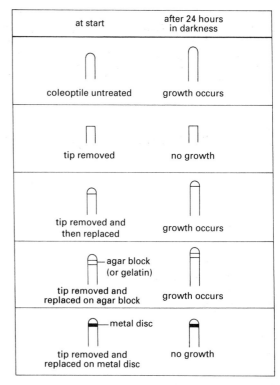

Figure 27.4 Coleoptile experiments

- the shoot tip is essential for growth and produces a chemical;
- this chemical messenger diffuses down to lower regions of the shoot where it stimulates growth by making cells elongate;
- The growth substance is able to diffuse through agar (or gelatin) but not through metal.

The hormone involved is now known to be one of a group of plant growth substances called **auxins**. The most common auxin is **indole acetic acid (IAA)**.

Only a low concentration of auxin is required to produce an effect. It promotes cell elongation by increasing the plasticity of the cell wall enabling it to stretch when water enters by osmosis during vacuolation.

Sites of IAA production

The auxin IAA is produced by the root and shoot tips and leaf meristems of plants. It is transported away from the tips. Movement of auxin over short distances, for example from cell to cell, occurs by diffusion. Long distance transport of auxin occurs in the plant's phloem tissue.

Auxins affect several different aspects of plant growth as follows.

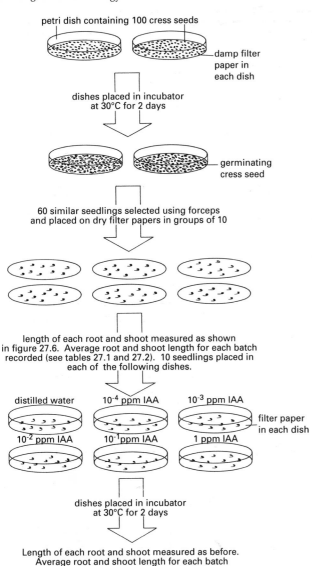

Figure 27.5 Investigating effects of IAA on roots and shoots

Investigating the effects of IAA on roots and shoots

The experiment shown in figure 27.5 is set up to investigate the effects of specific concentrations of IAA on the growth of roots and shoots of cress seedlings.

The length of each root and shoot is measured before and after contact with IAA by placing it on a glass slide lying on graph paper as shown in figure 27.6.

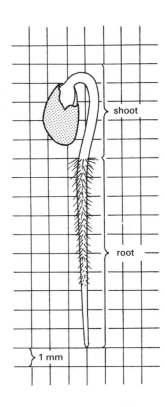

Figure 27.6 Cress seedling

	IAA concentration (ppm)					
	0 (control)	**10^{-4}**	**10^{-3}**	**10^{-2}**	**10^{-1}**	**1**
average length of root before IAA treatment (mm)	10	10	10	10	10	10
average length of root after IAA treatment (mm)	20	22	21	19	15	11
average increase in length (mm)	10	12	11	9	5	1
average change in length relative to control (mm)	0	2	1	−1	−5	−9
% stimulation (+) or inhibition (−) of growth	0	+20	+10	−10	−50	−90

Table 27.1 Root results

	IAA concentration (ppm)					
	0 (control)	10⁻⁴	10⁻³	10⁻²	10⁻¹	1
average length of shoot before IAA treatment (mm)	5	5	5	5	5	5
average length of shoot after IAA treatment (mm)	10	10	10.2	11	13	18
average increase in length (mm)	5	5	5.2	6	8	13
average change in length relative to control (mm)	0	0	0.2	1	3	8
% stimulation (+) or inhibition (−) of growth	0	0	+4	+20	+60	+160

Table 27.2   Shoot results

The solutions of IAA are made up by dissolving 0.1 g of IAA in 100 ml of water to give a stock solution containing 1000 parts per million (ppm) and then further diluting this to produce the five concentrations of IAA required. (These are kept in darkness when not in use.)

Tables 27.1 and 27.2 give typical sets of results. Since the seedlings in distilled water act as the control, an average increase in length greater than that found in water is regarded as a measure of **stimulation** of growth. An average increase in length less than that in water is regarded as **inhibition** of growth.

The effect of IAA concentration on a root or shoot is therefore quantified by subtracting the average increase in length of the control from average increase in length due to IAA (where negative values indicate inhibition).

Each result is then expressed as a percentage stimulation or inhibition of growth using the formula:

$$\frac{\text{average change in length relative to control}}{\text{average increase in length of control}} \times \frac{100}{1}$$

Figures 27.7 and 27.8 show bar graphs of the results. Under ideal circumstances, and using an even wider range of IAA concentrations, results are obtained which can be presented as two line graphs (see figure 27.9).

From these results it is concluded that very low concentrations of auxin which stimulate elongation of roots have little or no effect on shoots. On the other hand, higher concentrations of auxin which stimulate shoots inhibit the elongation of roots.

Table 27.3 lists the reasons for adopting certain techniques and precautions during the experiment.

Growth curvature effects

The experiment shown in figure 27.10 shows that auxin will diffuse from a shoot tip into an agar

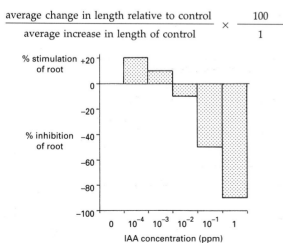

Figure 27.7   Bar graph of root results

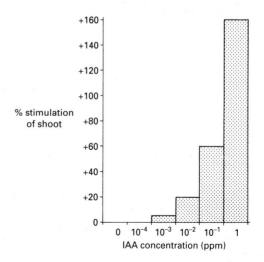

Figure 27.8   Bar graph of shoot results

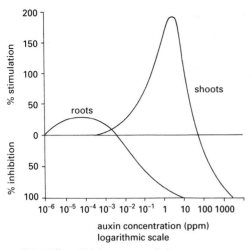

Figure 27.9   Effect of IAA on roots and shoots

design feature or precaution	reason
200 seeds germinated at start	to allow 60 similar seedlings to be selected
large number (10) seedlings used in each dish	to make results statistically valid
IAA solutions kept in dark	to prevent light destroying IAA
separate syringes used to add IAA solutions to dishes	to avoid contamination of one IAA concentration by another
all factors kept equal except IAA concentration	to ensure that IAA concentration is only variable factor involved in experiment

Table 27.3   Design techniques

block and then from the agar into a cut coleoptile where the cells resume elongation.

When an agar block containing auxin is placed assymetrically on a decapitated coleoptile (see figure 27.11), the shoot bends because the side below the agar block received a higher

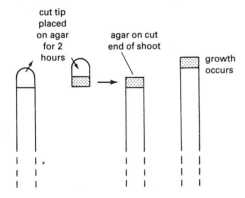

Figure 27.10   Use of agar block

concentration of growth substance causing greater cell elongation on that side.

Within limits (see graph in figure 27.9) the *higher the concentration* of auxin, the *greater the curvature* produced.

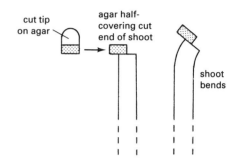

Figure 27.11   Bending effect

Phototropism

This is the name given to a **directional growth movement** by a plant organ in response to **light** from one direction.

Oat coleoptiles exhibit **positive phototropism** by bending towards a unidirectional source of light as shown by shoot **X** in figure 27.12. Since shoot **Y** fails to bend, it is concluded that the shoot tip is the region responsible for detecting unilateral light.

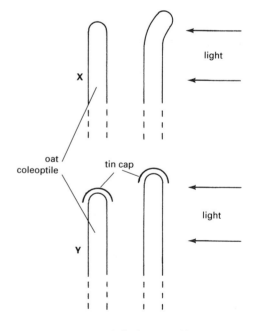

Figure 27.12   Response to light from one side

Mechanism of phototropism

The experiment shown in figure 27.13 shows that a unidirectional light source causes an *unequal* distribution of hormone to occur in the shoot tip.

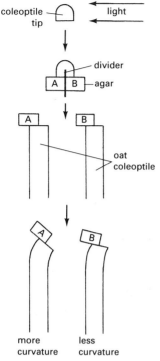

Figure 27.13   Mechanism of phototropism

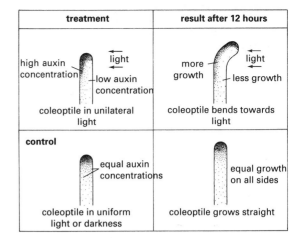

treatment	result after 12 hours
high auxin concentration / low auxin concentration — coleoptile in unilateral light	more growth / less growth — coleoptile bends towards light
control equal auxin concentrations — coleoptile in uniform light or darkness	equal growth on all sides — coleoptile grows straight

Figure 27.14   Hormonal explanation of phototropism

A higher concentration of **auxin** is present in the *non-illuminated* side than in the illuminated side.

These findings can be used to explain the mechanism of positive phototropism. If the shaded side of a shoot contains more auxin, then more cell elongation will occur on that side. This makes the shoot bend towards the light as shown in figure 27.14.

Geotropism

This is the name given to a **directional growth movement** by a plant organ in response to **gravity**.

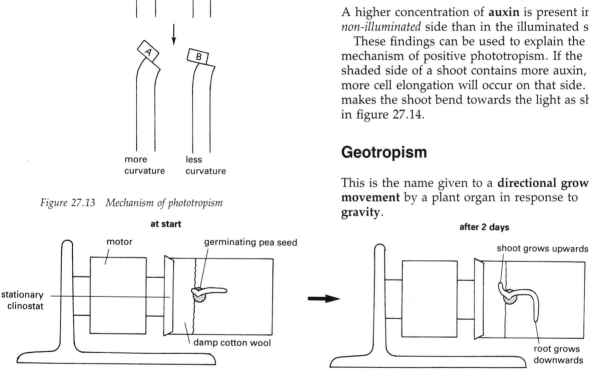
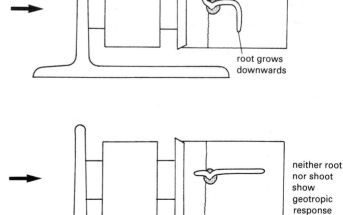

Figure 27.15   Geotropism

The experiment shown in figure 27.15 shows the effect of gravity on the growth of a root and a shoot.

In the stationary clinostat, the root exhibits **positive geotropism** by growing downwards towards the source of the stimulus (gravity). The shoot shows **negative geotropism** by growing upwards away from the pull of gravity.

In the rotating clinostat (the **control** experiment), gravity acts equally on all sides of each plant organ and therefore no geotropic responses occur.

Mechanism of geotropism

In a shoot, it is thought that a high concentration of auxin gathers and promotes cell elongation on the lower surface. This causes the shoot to bend and grow upwards, i.e. negative geotropism (see figure 27.16).

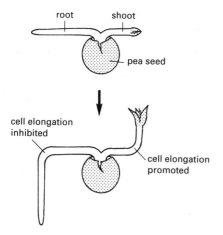

Figure 27.16   Mechanism of geotropism

In a root it is thought that an **inhibitor** substance (not auxin) slows down cell elongation on the lower surface causing the root to bend and grow downwards i.e. positive geotropism.

Apical dominance

High concentrations of auxin inhibit growth (see figure 27.9).In many plants, the apical (terminal) bud is able to inhibit the development of lateral (side) buds further down the stem by producing a sufficiently high concentration of auxin which is translocated down the stem's phloem tissue to the side buds.

Such auxin-controlled **apical dominance** is demonstrated by the experiment shown in figure

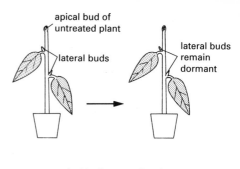

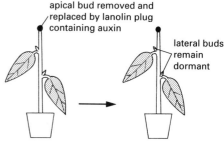

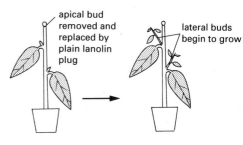

Figure 27.17   Apical dominance

27.17. In the absence of the apical bud (or an auxin-containing substitute), hormone concentration drops to a level which no longer inhibits growth of lateral buds. These now develop into side branches.

Leaf abscission

Abscission is the separation of leaves (and fruit) from a plant. Prior to leaf fall, auxin concentration drops and a thin abscission layer of cells is formed at the base of the leaf stalk (petiole) as shown in figure 27.18.

The walls of the cells in this layer gradually become weakened. The leaf stalk eventually snaps and the leaf falls.

During the growing season, auxin prevents the abscission layer from forming and so inhibits abscission of leaves (and fruit).

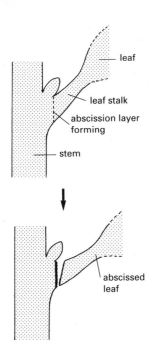

Figure 27.18   Leaf abscission

A	Result of experiment considered: 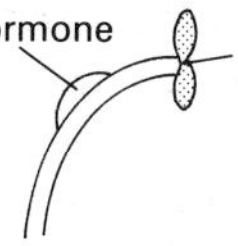	hormone — shoot in uniform light bends away from hormone
B	Prediction made: 'application of hormone to one side of a shoot in uniform light will cause it to bend away from the side treated with hormone.'	
C	Conclusion drawn: 'hypothesis is supported.'	
D	Prediction tested by experiment: 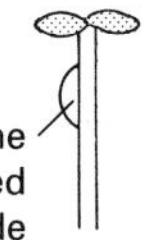	hormone applied to one side
E	Hypothesis constructed: 'bending caused by presence of higher concentration of hormone on non-illuminated side of shoot.'	
F	Observation made: 'shoot bends towards light coming from one direction.' 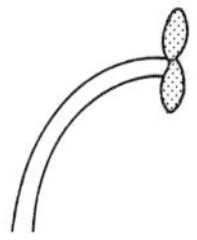	light from one direction

Figure 27.19

Fruit formation

Following fertilisation in a flower, auxin promotes the formation of the fruit coat from the ovary wall. The coat may become soft and succulent, as in a grape for example.

Fruit development without fertilisation (called **parthenocarpy**) can be induced artificially by treating unfertilised flowers with auxin. This process produces a plentiful supply of seedless fruit that is ripe for harvesting at the same time.

QUESTIONS

1 a) Name the growth hormone made in the human body.
   b) Suggest why hormones are described as chemical messengers.
   c) State TWO differences between acromegaly and pituitary dwarfism in humans.
2 Figure 27.19 shows six stages involved in constructing and testing a hypothesis. Arrange the stages into the correct order starting with **F**.
3 a) Predict the direction in which the young root in figure 27.20 will grow. Explain your answer.
   b) The young shoot was found to grow

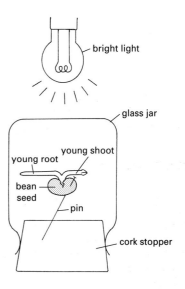

Figure 27.20

vertically upwards. A pupil concluded therefore that the shoot was showing a negatively geotropic response. Why was she not justified in drawing this conclusion?
c) How would the experiment have to be altered to find out if the shoot is able to make a negatively geotropic response?
4 Leaves **X** and **Y** in figure 27.21 (i) were approaching an age at which they would be lost naturally from each plant by abscission. One of these leaves was then treated with auxin.
Figure 27.21 (ii) shows the appearance of the plants three weeks later.
a) Which leaf was treated with auxin?
b) Give TWO reasons for your answer.

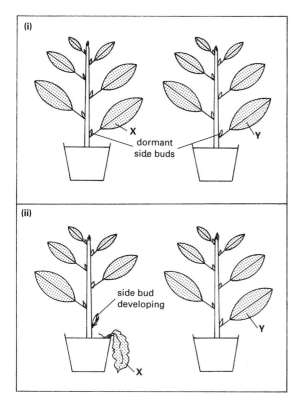

Figure 27.21

EXPERIMENTAL DESIGN QUESTION

Observations: A certain brand of rooting powder (RP) which contains auxin promotes growth of cuttings taken from hardwood plants such as roses.

Cuttings dipped in water and then in RP pick up more powder and 'take' more easily than cuttings dipped only in RP before planting.

Hypothesis: Water-dipped cuttings grow better because a higher concentration of auxin stimulates a greater amount of root formation.

Problem: Design an experiment which would enable you to test the validity of the hypothesis above. (Assume that RP is water soluble.)

Your design should include reference to:
● relevant control(s),
● measurements needed to provide valid data,
● method of calculating percentage stimulation of root formation.

28 Hormonal influences on growth – part 2

Gibberellins

In the 1920s, a Japanese farmer discovered that some of his rice plants which were infested with a fungus had grown abnormally tall. This growth effect was found to have been caused by a growth substance (hormone) secreted by the fungus. Since the fungus was called *Gibberella*, the hormone was called **gibberellin**.

Many gibberellins are now known to exist. They make up a second group of plant growth substances which occur naturally in small amounts in most plants. The most common one is **gibberellic acid (GA)**.

Like auxins, gibberellins stimulate cell division and elongation in stems. However they play no part in tropic movements or bending of shoots. Gibberellins affect several other aspects of growth and development as follows.

Effect of gibberellic acid (GA) on dwarf pea seedlings

Look at the experiment shown in figure 28.1. The dwarf pea seedlings in pot **A** which receive gibberellic acid reach the same height as the normal tall variety in pot **C**. This is due to an

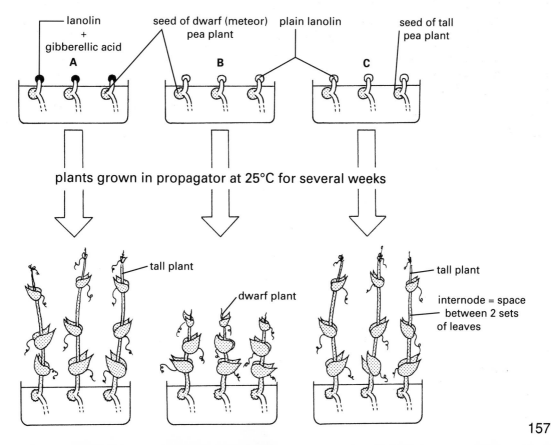

increase in length (not number) of their internodes.

The dwarf plants in the control pot **B** show that this effect is not due to the application of lanolin alone.

It is concluded that gibberellin is needed by a plant to make its internodes increase in length. Tall varieties are able to manufacture sufficient gibberellin to make them grow to full height. The dwarf condition (which is inherited) is caused by a shortage of gibberellin due to a genetic deficiency.

Effect of gibberellic acid (GA) on germinating barley grains

Figure 28.2 shows the internal structure of a soaked barley grain. It indicates the plane in which the grain should be cut to give an **embryo** 'half' and an **endosperm** 'half'. Several of these 'halves' are needed for experiment shown in figure 28.3.

The results of this experiment show that digestion of starch (in the starch agar) has only occurred in plate **A** beneath the endosperm parts of barley grains.

It is concluded that both endosperm and gibberellic acid must be present for the starch-digesting enzyme, **α-amylase**, to be produced in a barley grain.

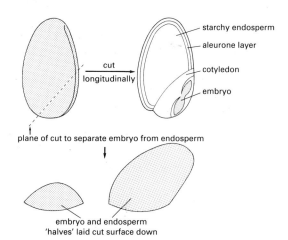

Figure 28.2 Barley embryo and endosperm

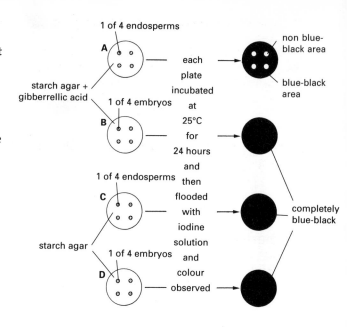

Figure 28.3 Effect of GA on barley endosperm

Under normal circumstances in a soaked cereal grain, the hormone gibberellin is made by the embryo and passes up to the **aleurone** layer (see figure 28.4). Here it acts at gene level and induces the production of α-amylase. This digests starch to sugar (maltose) which is needed for growth by the seedling.

By this means, gibberellin breaks the dormancy of many types of seed and induces them to germinate.

Model of hormone action

By definition, a hormone is a chemical substance produced at one site and transported to another

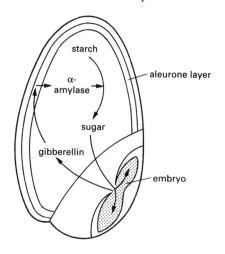

Figure 28.4 Induction of α-amylase by GA

site where a low concentration of it brings about an effect.

Gibberellin is produced at one site (the embryo) and is active at another site (the aleurone layer) where, in low concentration, it exerts its effect (induction of α-amylase). Hence this sequence of events provides a model of hormone action.

Effect of gibberellic acid (GA) on bud dormancy

The buds of deciduous trees remain dormant during winter. Under natural conditions in spring, GA produced by the plant breaks this dormancy and allows the buds to open and develop into new leafy branches.

If GA is applied to buds during the winter or early spring, their dormancy is broken artificially and they begin to develop.

Applications of growth substances

Herbicides

Synthetic auxins are used as **selective weedkillers** (herbicides). They work by stimulating the plant's rate of growth and metabolism to such an extent that the plant exhausts its food reserves and dies of starvation.

Selective weedkillers are especially effective on lawns since weeds with broad leaves (see figure 28.5) absorb much of the chemical and die whereas narrow-leaved grass plants absorb little of the chemical and are hardly affected.

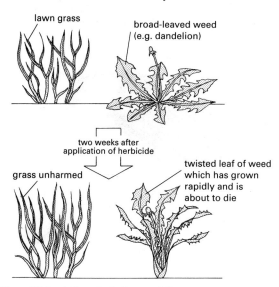

Figure 28.5 Action of selective weedkiller

Rooting powders

Roots which arise from a plant part other than the main root (or its branches) are called **adventitious** roots.

When rooting powder which contains synthetic auxin is applied to the cut ends of stems, it stimulates the formation of adventitious roots. This makes the plant easy to propagate.

QUESTIONS

1 a) (i) What effect does gibberellic acid (GA) have on stem length when applied to the shoots of dwarf pea plants?
(ii) Is this effect due to increase in *number* or *length* of internodes?
b) (i) The action of GA in a germinating cereal grain is said to provide a model of hormone action since GA is produced at one site and is active at another. Identify these two sites and describe the activity which occurs at the second one.
(ii) If the experiment shown in figure 28.3 is not properly set up then a small amount of digestion may also be found to occur in a plate other than **A**. Suggest a possible source of error responsible for this result.
2 Most cereal crops are narrow-leaved; the weeds that compete with them tend to be broad-leaved.
a) What type of chemical could be used in an attempt to destroy only the weeds?
b) Explain how this method works.
3 Copy and complete table 28.1 by placing ticks in the appropriate columns to indicate the roles played by gibberellin (G) and auxin (A).

role	G only	A only	G and A
induction of α-amylase production in cereal grains	✓		
stimulation of cell division and elongation			✓
reversal of genetic dwarfism	✓		
stimulation of adventitious root formation		✓	
prevention of leaf abscission		✓	
promotion of phototropic growth movements		✓	

Table 28.1

DATA INTERPRETATION

In an experiment, four groups of ten French bean plants were treated as shown in table 28.2.

The lengths of each plant's side buds and shoots were measured at the start and at 2-day intervals. The average total length of side buds and shoots per plant was calculated for each group and recorded in table 28.3.

a) Draw line graphs of the data on the same sheet of graph paper. (2)

b) (i) In which group of plants did apical dominance occur?

(ii) Explain how you arrived at your answer. (2)

c) What is the purpose of including group Y in the experiment? (1)

d) From the data it was concluded that application of one of the hormones stimulated growth of side shoots.

(i) Identify the hormone.

(ii) Explain why the data justifies drawing such a conclusion.

(iii) Calculate the percentage stimulation brought about by this hormone at day 8. (3)

e) Suggest why the result recorded for group X at day 10 differs from earlier results recorded for the same group. (1)

f) Why were ten plants used for each treatment? (1)

(10)

group	treatment
W	apical buds removed and no further treatment given
X	apical buds replaced with lanolin plugs containing auxin
Y	apical buds replaced with plain lanolin plugs
Z	apical buds replaced with lanolin plugs containing gibberellin

Table 28.2

		time in days from start of experiment					
		start	2	4	6	8	10
average total length of side buds and shoots per plant (mm)	W	4	4	10	35	60	96
	X	4	4	4	4	6	20
	Y	4	4	11	34	60	98
	Z	4	4	16	56	102	146

Table 28.3

What you should know (CHAPTERS 27–28)

1 The **pituitary gland** produces both **somatotrophin** (which promotes human growth) and **thyroid-stimulating** hormone (TSH).

2 The **thyroid** gland responds to TSH by producing **thyroxine** which controls metabolic processes.

3 The **auxins** make up a group of plant growth substances (hormones).

4 Auxin is produced at root and shoot tips and promotes **cell elongation**.

5 **Low** concentrations of auxin which stimulate **root** elongation have little effect on shoots. **Higher** concentrations of auxin which stimulate **shoot** elongation **inhibit** root elongation.

6 **Unequal** distribution of auxin causes shoot to **bend** producing **tropic** movements.

7 Apical buds exert **apical dominance** by producing high concentrations of auxin which inhibit growth of side buds.

8 Auxin prevents **leaf abscission** and promotes fruit formation.

9 Synthetic auxins act as **herbicides** by disrupting growth of broad-leaved weeds more than grass-like plants which absorb less through their narrow leaves.

10 Synthetic auxins stimulate **root formation** on stem cuttings.

11 The **gibberellins** make up a second group of plant growth substances.

12 Gibberellin promotes **cell division** and **elongation** in stems.

13 Gibberellin **reverses genetic dwarfism** by promoting stem internode elongation.

14 Gibberellin **induces production** of α-**amylase** in cereal grains and breaks dormancy in winter buds.

29 Effect of chemicals on growth

Macro-elements in plants

In addition to the large amounts of carbon (C), hydrogen (H) and oxygen (O) needed for the manufacture of carbohydrates during photosynthesis, a green plant requires small but appreciable quantities of several other chemical elements for healthy growth.

Four of these **macro-elements** are **nitrogen** (N), **phosphorus** (P), **potassium** (K), and **magnesium** (Mg). They are present in fertile soil and are absorbed by the plant's root hairs.

Table 29.1 shows the results of an experiment in which three different species of plant were grown in the same soil under identical environmental conditions. When analysed, all three species were found to contain significant, though not identical, amounts of each macro-element.

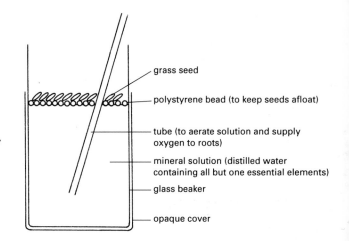

Figure 29.1   Water culture experiment

	percentage of dry weight			
	N	P	K	Mg
barley	1.94	0.13	4.04	0.29
sunflower	1.47	0.08	3.47	0.73
wheat	2.26	0.06	4.16	0.23

Table 29.1   Macro-element content of plants

Water culture experiments

The importance of each macro-element to a plant can be investigated by setting up a series of **water culture** experiments where one element at a time is omitted from the mineral solution. The control experiment contains all elements necessary for growth.

The glass beaker used in the water culture experiment in figure 29.1 is initially rinsed with concentrated nitric acid to remove traces of mineral elements. It is surrounded by an opaque cover to keep out light and prevent growth of

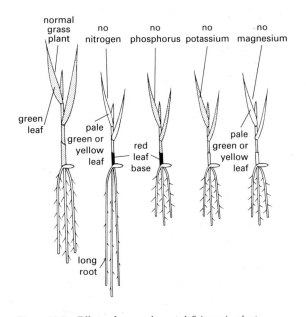

Figure 29.2   Effects of macro-element deficiency in plants

algae. (If these were allowed to grow, a second variable factor would be introduced into the experiment making the results invalid.)

After several weeks' growth, each plant lacking an essential macro-element is found to display certain **deficiency symptoms** when compared with the control (see figure 29.2). These symptoms often indicate the role normally played by the macro-element in the plant's metabolism as summarised in table 29.2.

element omitted	symptoms of deficiency	reason for deficiency symptom (role of element)
nitrogen	overall growth reduced, leaves **chlorotic** (pale green or yellow), leaf bases red, roots long and thin	required for formation of protein and nucleic acids
phosphorus	overall growth reduced, leaf bases red	required for formation of ATP and nucleic acids
potassium	overall growth reduced, early death of older leaves	required for essential role in protein synthesis
magnesium	overall growth reduced, leaves **chlorotic**	required for chlorophyll formation

Table 29.2 Effects of macro-element deficiency in plants

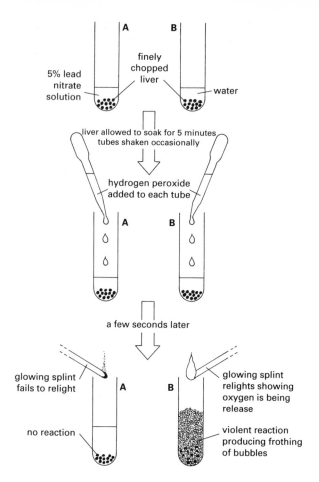

Figure 29.3 Effect of lead on catalase activity

Lead

Investigating the effect of lead nitrate on enzyme activity

Catalase is an enzyme made by living cells (e.g. liver). It catalyses the breakdown of poisonous hydrogen peroxide to water and oxygen:

$$2H_2O_2 \xrightarrow{\text{catalase}} 2H_2O + O_2 \uparrow$$

Look at the experiment shown in figure 29.3. In tube **B** (the control) oxygen is released showing that catalase is active.

However, in tube **A**, no reaction occurs. This shows that the activity of catalase has been *inhibited* by lead nitrate.

Harmful effects of lead

'Pica' is the name given to the habit of eating non-foodstuffs. If these include flakes of peeling paint containing lead, and the consumers are very young children, then **lead poisoning** results. At first the symptoms include vomiting, clumsiness, irritability and headaches but if massive intakes of lead continue then the outcome is drastic and irreversible.

Certain brain cells are directly injured by lead. In addition, capillary walls become too permeable; they leak, causing the brain to swell. Since the brain is encased in the rigid skull, some of its tissues become damaged. The effects on the young children involved can range from subtle learning difficulties to profound mental retardation.

These findings apply solely to a minority of unfortunate young children. They raise an important question: can a level of lead absorption that is insufficient to cause obvious acute symptoms nevertheless cause small amounts of undetected brain damage?

It is a fact that the air breathed by the residents of densely populated cities is rich in lead released in exhaust **fumes** from petrol engines. Drinking water carried to households in **lead pipes** may also contain unacceptable levels of dissolved lead salts.

A detailed research programme was carried out in USA to assess whether these forms of milder lead poisoning have less obvious but still harmful, effects.

Lead which has gained access to the body is stored in bones and teeth. An accurate measurement of total exposure to lead can therefore be obtained by analysing tooth samples for their lead content. This was done for 3000 young children. The children were also given a range of intelligence tests and assessed by their teachers on various aspects of behaviour and classwork.

Figure 29.4 shows a correlation between high lead content of teeth and factors contributing to poorer overall performance. In addition 'high lead' children were found, on average, to be four points behind the others in intelligence tests.

Steps are now being taken to remove lead from petrol and replace lead pipes with copper.

Iron and calcium in animals

Unlike lead, **iron** and **calcium** are essential elements needed for healthy growth and development.

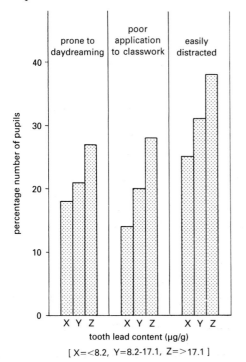

Figure 29.4 Effects of lead on schoolchildren

Iron

This element is an essential constituent of many **enzymes** (e.g. catalase). It is needed to make **cytochrome** (the hydrogen carrier molecule in respiration, see page 22) and it forms part of the haem group in the respiratory pigment **haemoglobin** (see page 44).

A small amount of iron is lost every day by humans in dead skin cells, bile and urine. It therefore needs to be replaced. Growth and menstruation make further demands as shown in table 29.3.

factor creating demand for iron	average amount of iron required to satisfy demand (mg/day)
basic loss in skin cells, bile and urine	0.5–1.0
growth: 0–1 year	0.7–0.8
1–11 years	0.3
adolescence	0.5
normal menstruation	0.5–1.6
pregnancy	6.0

Table 29.3 Iron demands

It is therefore essential that growing children and pregnant women consume a diet containing large amounts of iron. Many foodstuffs such as meat, eggs, fish, green vegetables and cereals are rich in iron. However iron present in vegetables and cereals is less well absorbed than that from animal sources. Vegetarians must be careful in their choice of diet if they are to avoid iron deficiency and anaemia.

Calcium

This element is required by animals to form calcium salts such as **calcium phosphate** which becomes built into the hard parts of **bones** and **teeth**. Calcium carbonate is required by many invertebrates to form their shells.

In humans, calcium is also needed for the **clotting of blood** and the **contraction of muscle**. The main sources of calcium are milk, cheese and green vegetables.

Vitamin D deficiency

Vitamin D is essential in humans to promote the absorption of **calcium** and **phosphate** from the intestine and their uptake by bone.

Deficiency of vitamin D in young children leads to the formation of soft abnormal bones. The long bones of the legs tend to bend under the weight of the body making the child bow-legged. This condition is known as **rickets**. It is prevented by consuming foods such as cod liver oil, egg yolk or full cream milk which are rich in vitamin D.

Rickets is almost unknown in tropical countries because vitamin D is formed in the skin when it is exposed to the ultraviolet rays present in sunshine. The condition is more likely to be found amongst poorly fed children in cities of northern latitudes where sufficient ultraviolet radiation fails to penetrate the cloud cover and layer of atmospheric pollution.

Effect of drugs on foetal development

Thalidomide

In the 1950s this tranquilliser was developed and administered to pregnant women in order to counteract the feelings of nausea which many experience. If it was taken during a certain critical period in very early pregnancy then the **limbs** of the foetus failed to develop properly.

As a result the baby's hands developed attached to the shoulders and the feet to the hip joints. In addition malformation of eyes, ears and heart occurred in some cases together with mental subnormality and epilepsy.

This tragedy demonstrated the danger of taking drugs during pregnancy. Thalidomide has now been withdrawn from use.

Alcohol

Alcohol contains energy but its value as a food is restricted by its depressant action on the central nervous system. When consumed in excess, alcohol has a **toxic effect** on the liver.

A higher incidence of spontaneous abortion is found to occur amongst women who drink in excess during pregnancy. In those women who do not suffer a miscarriage but who do continue to drink heavily, alcohol crosses the placenta and causes the blood vessels in the umbilical cord to collapse temporarily (see figure 29.5). During these periods the foetus fails to receive an adequate **oxygen supply** vital for the proper development of growing tissues such as the brain.

Alcohol also interferes with the normal absorption of nutrients such as vitamin B6 and zinc from the mother's gut. These are essential for the health of both the mother and the developing embryo.

In extreme cases of alcohol abuse the foetus suffers several harmful effects known collectively as the **foetal alcohol syndrome**. These include:
- pre- and post-natal growth retardation;
- facial abnormalities;
- heart defects;
- development of abnormal joints and limbs;
- mental retardation.

To eliminate the possibility of even the mildest form of damage being caused to the foetus, women are advised to completely avoid drinking alcohol during pregnancy.

Alcohol abuse by men has a harmful effect on sperm production. A higher incidence of abnormalities is therefore found to occur amongst the offspring of chronic alcoholics married to non-drinking women.

Nicotine

Nicotine is a poisonous drug present in tobacco plants. It stimulates the central nervous system and increases pulse rate and blood pressure. In addition to nicotine, tobacco smoke contains **carbon monoxide (CO)** and **tar** (which has been shown to contain up to 30 different harmful substances). A higher incidence of lung cancer and coronary artery disease is found to occur amongst heavy smokers.

Women who smoke during pregnancy retard the growth of their unborn children. CO reduces the amount of **oxygen** that the blood can carry and nicotine in the bloodstream prevents adequate **glucose** reaching foetal tissues including brain cells.

Babies born to smokers are therefore smaller than average and have low blood sugar levels. Evidence also suggests that they do not develop intellectually at the same rate as children of non-smokers. Foetal development is also adversely affected if the mother, herself a non-smoker, lives with smokers and is constantly subjected to a smoky environment.

QUESTIONS

1 **a)** Name an element which, if in deficient supply to a plant, brings about chlorosis.
b) Relate this element's role in the plant's metabolism to chlorosis.
c) Chlorosis rarely becomes apparent until a few weeks after germination. Suggest why.
d) What control should be set up in a water culture experiment?

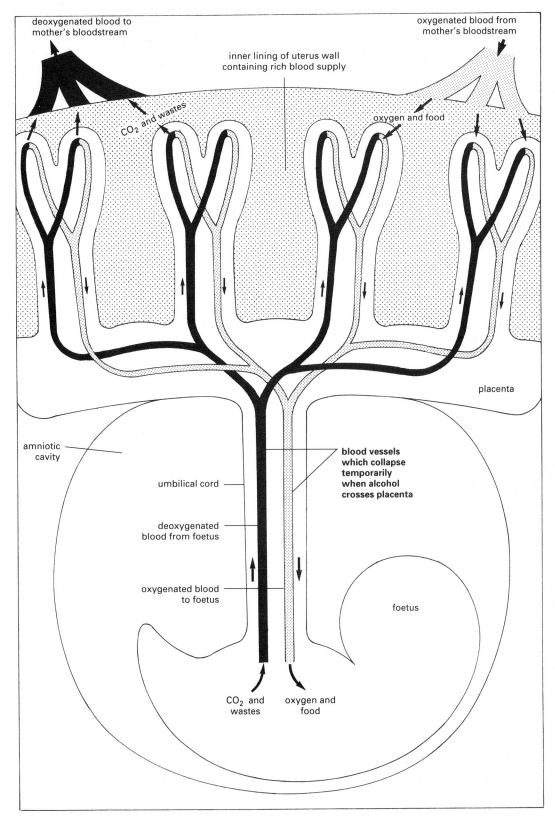

deoxygenated blood to
mother's bloodstream

oxygenated blood from
mother's bloodstream

inner lining of uterus wall
containing rich blood supply

CO_2 and wastes

oxygen and food

placenta

amniotic
cavity

blood vessels
which collapse
temporarily
when alcohol
crosses placenta

umbilical cord

deoxygenated
blood from foetus

oxygenated blood
to foetus

foetus

CO_2 and
wastes

oxygen and
food

Figure 29.5 Umbilical cord blood vessels

2 Although the iron content of porridge oats is 4 mg/100 g, only 10% of this iron is absorbed from the gut into the bloodstream.

a) If a woman in late pregnancy needs a daily uptake of 6 mg of iron, how much porridge would she have to eat to meet this iron demand?

b) Since this would be unrealistic, name THREE other foods that she could also consume in order to gain iron.

3 Figure 29.6 shows the daily turnover of calcium in a human adult who has consumed 900 mg. Bone is continually being remodelled by the processes of bone formation and resorption. These processes are equal in an adult whose bones have reached their final size.

a) State the figure that should be inserted in boxes X and Y.

b) Assume that a 13-year-old girl also consumes 900 mg of calcium each day. Suggest a way in which her overall calcium turnover would differ from that of an adult.

4 During the 1960s, many children of Asian immigrants who had settled in British cities were found to be suffering from rickets.

a) What vitamin must have been deficient in the children's diet?

b) Why did the children's parents, who had been fed a similar diet during their childhood in Asia, not suffer from rickets in their youth?

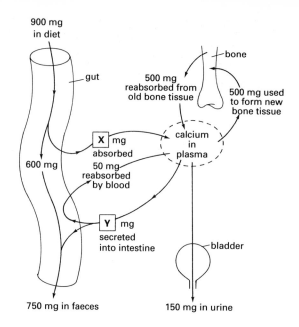

Figure 29.6

c) Describe an obvious physical feature displayed by a 2-year-old child suffering rickets.

d) Name a food rich in the vitamin needed to cure rickets.

e) Briefly describe the role played by this vitamin in the body.

30 Effect of light on growth

Vegetative shoot

A **vegetative** shoot is one which does not bear flowers. To investigate the effect of light on growth and development of a leafy shoot, two sets of broad bean seedlings are grown in

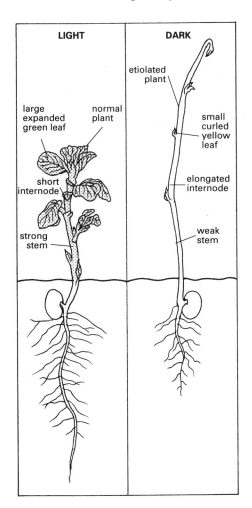

Figure 30.1   Etiolation

identical environmental conditions except that one is kept in **daylight** and the other in **darkness**.

From a comparison of the results (see figure 30.1) it is concluded that light is necessary for the development of fully **expanded green leaves** and **short strong stem internodes**. (An internode is a region of stem lacking leaves and branches.)

Plants grown in darkness and possessing **small curled yellow leaves** and **long weak internodes**, are said to be **etiolated**. Microscopic examination reveals that increased cell elongation accounts for the formation of these long internodes. It is thought that the cells of an illuminated plant do not become as elongated because they receive less auxin (since some has been destroyed by light).

In terms of evolution, etiolation is of survival value to a plant since the chance of some leaves eventually reaching light is increased by the plant expending most of its resources on the development of long thin stems.

Effect of light on flowering

Some plants stop producing vegetative (leaf) buds and start producing flower buds in response to a change in the period of illumination (**photoperiod**) to which they are exposed. Such a response to a photoperiod is called **photoperiodism**.) Three distinct types of plant exist.

Long day (short night) plants

These plants only flower when the number of hours of light to which they are exposed is **above** a certain critical level (i.e. the number of hours of darkness is below the critical level).

In spinach (*Spinacea oleracea*), shown in figure 30.2, the plant must receive at least 13 hours of light in order to flower.

The length of the critical period of illumination

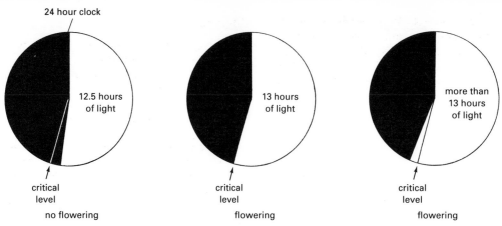

Figure 30.2   *Photoperiodism in spinach, a long day (short night) plant*

species of long day (short night) plant	critical duration of light (hours)	species of short day (long night) plant	critical duration of darkness (hours)
Dill	11	*Bryophyllum*	12
Italian ryegrass	11	*Chrysanthemum*	9
Red clover	12	Cocklebur	9
Spinach	13	Strawberry	14
Winter wheat	12	Winter rye	12

Table 30.1   *Critical periods of light and darkness*

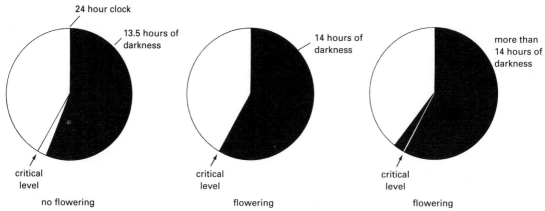

Figure 30.3   *Photoperiodism in strawberry, a short day (long night) plant*

varies amongst different species of long day (short night) plants (see table 30.1). In some cases it is found to be less than 12 hours.

Short day (long night) plants

These plants only flower when the number of hours of light to which they are exposed is **below** a certain critical level (i.e. the number of hours of darkness is above the critical level).

In strawberry (*Fragaria chiloensis*), shown in figure 30.3, the plant must receive at least 14 hours of darkness in order to flower.

The duration of the critical period of darkness varies amongst different species of short day (long night) plants (see table 30.1). In some cases it is found to be less than 12 hours.

Day neutral plants

These are plants in which flowering is not dependent upon photoperiod. Examples include celery, geranium, snapdragon and tomato.

Mechanism of the response

Plants contain tiny amounts of a light-sensitive pigment called **phytochrome** which exists in two forms. **Phytochrome 660 (P_{660})** absorbs red light (with an absorption peak at a wavelength of 660 nm) and **phytochrome 730 (P_{730})** absorbs light in the far red (non-visible) region of the spectrum (with an absorption peak at wavelength 730 nm) as shown in figure 30.4.

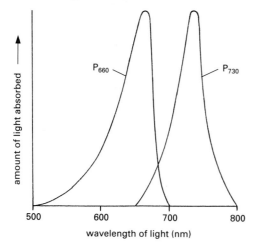

Figure 30.4   Absorption spectra for phytochromes

On absorbing its particular wavelength of light, each form of phytochrome becomes converted into the other form (see figure 30.5).

Since sunlight contains much more light of wavelength 660 nm than 730 nm, during daylight P_{660} becomes converted to P_{730} which then accumulates. P_{730} is unstable and during darkness slowly reverts to P_{660} which accumulates as shown in figure 30.6.

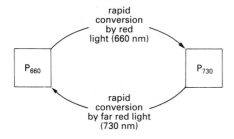

Figure 30.5   Interconversion of phytochrome

Figure 30.6   Interconversion of phytochrome

Florigen

Flowering in plants is thought to be initiated by a hormone (or combination of hormones). Since this hormone has not yet been isolated, it is provisionally referred to as '**florigen**'.

In long day plants, a high concentration of P_{730} is required for the release of florigen (see figure 30.7).

In short day plants, a high concentration of P_{660} is required for the release of florigen (see figure 30.8). If the long night is interrupted somewhere near its middle by a flash of red light, flowering does not occur because some of the essential P_{660} has reverted to P_{730}.

The experiment shown in figure 30.9 shows that it is the *leaf* that perceives the stimulus and

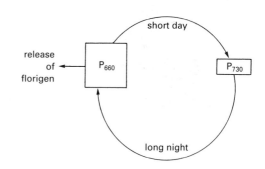

Figure 30.7 P_{730} formation in long day plant

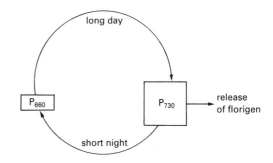

Figure 30.8 P_{660} formation in short day plant

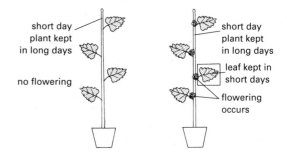

Figure 30.9   Site of stimulus perception

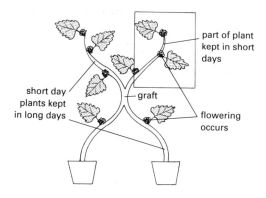

Figure 30.10   Transport of florigen

produces sufficient florigen to induce flowering in the entire plant.

The experiment in figure 30.10 shows that florigen is *translocated* in the *phloem* to meristems where flowering is induced by the switching on of certain genes.

Latitude and season

Photoperiodism affects the geographic distribution of plant species since daylength varies with **latitude** and **season** (see figures 30.11 and 30.12).

At the equator, 12-hour daylengths alternate with 12-hour periods of darkness through the

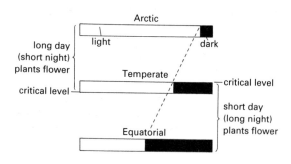

Figure 30.11   Different latitudes (north hemisphere) in July

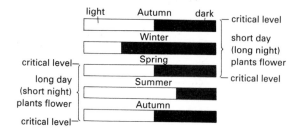

Figure 30.12   Different times of year in temperature latitude

whole year. However in other parts of the world, seasonal variations occur in the length of the daylight period. The further away from the equator a geographical location is situated, the longer its daylengths in summer and the shorter its daylengths in winter.

Short day plants (which need a critical duration of darkness) tend to live near the equator where they can flower all year or live in temperate regions where they flower from late autumn to early spring. They cannot reproduce in arctic environments because the temperature is too low for growth when the nights are long in winter.

Long day plants (which need a critical duration of light) tend to inhabit extreme northern (and southern) latitudes or temperate regions where they flower from late spring to early autumn when the daylengths are long.

Effects of light on gonadal activity

Behaviour in animals is described as **rhythmical** when it is repeated at definite intervals. Although such behaviour is **endogenous** (under internal control), the time at which it occurs is influenced by an external factor.

Many birds and mammals are **seasonal breeders**. Their gonads (testes and ovaries) become active only at a certain time of the year. These changes are triggered by the arrival of daily photoperiods of a certain critical length.

In birds, the reproductive activity of **long day breeders** is stimulated by the increasing daylengths that occur in spring. Evidence from experiments using the Japanese quail show that long photoperiods are perceived by **photoreceptors** in the hypothalamus of the bird's brain. This in turn stimulates the pituitary gland to secrete gonadotrophic hormones which promote the enlargement and activity of gonadal tissues and the production of sex cells as shown in figure 30.13. (In birds it is essential that the gonads remain very small outside the breeding season to keep the bird's mass to a minimum for flight).

Figure 30.13 also shows how a similar series of events occurs in many small mammals such as the rabbit which are long day breeders. Evidence suggests that in mammals but not in birds, the tiny **pineal gland** attached to the brain acts as the 'biological clock'. It is thought to receive the daily photoperiodic messages and convey the information to the brain. It does this by secreting the hormone **melatonin** in darkness only. This enables the brain to sense changes in the precise timing of the light/dark cycle (e.g. minimal

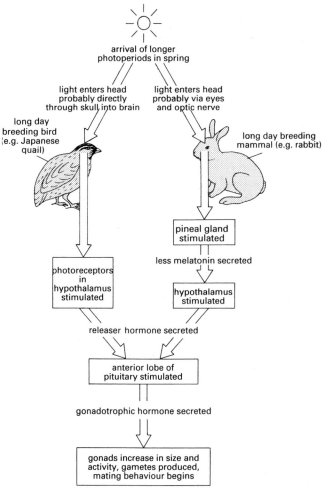

Figure 30.13 Effect of long photoperiods on gonadal activity

secretion of melatonin indicates long daylength).

In larger mammals (e.g. sheep) which require a longer period of gestation, a **short day breeding cycle** occurs. Gonadal activity and reproductive behaviour are triggered by the decreasing photoperiods which occur in autumn.

In each case, seasonal breeding behaviour is timed so that offspring will be born during *favourable environmental conditions*.

QUESTIONS

---

1 **a)** State TWO ways in which the (i) leaves (ii) stem internodes of an etiolated plant differ from those of a plant grown in light.
   b) Growth may be defined as (i) an increase in cell number, or (ii) an increase in dry mass. Briefly discuss whether or not an etiolated bean seedling shows real growth.
   c) Predict the ultimate fate of an etiolated bean seedling kept permanently in darkness. Explain your answer.
   d) With reference to (i) existing leafy shoots and (ii) new leafy shoots, predict the effect on transferring a healthy plant into total darkness for a few weeks.

2 Copy figure 30.14 which summarises photoperiodism in long and short day plants and complete the blanks using the following terms:
   daylight, release of florigen, P_{730}, darkness, production of flowers, P_{660}.

3 *Chrysanthemums* are short day plants which normally flower in autumn.
   a) Which part of the plant must be exposed to the critical photoperiod for flowering to occur?
   b) Describe the procedure that should be adopted by a horticulturist in order to produce a crop of *Chrysanthemum* plants in full bloom at Christmas time.

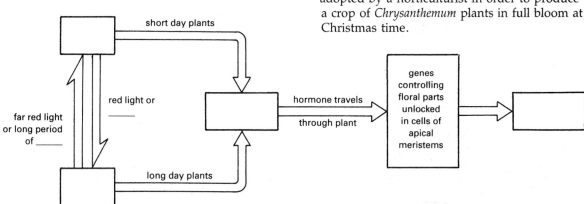

Figure 30.14

4 Which geographical locations tend to be inhabited by long day plants? Explain why.

5 The graph in figure 30.15 shows the effect of light on the production of gonadotrophic hormone made by male Japanese quail birds.

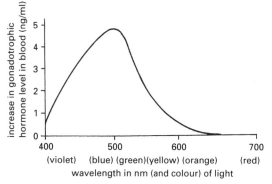

Figure 30.15

a) State clearly the variable factor controlled by the experimenter in this experiment.

b) Which light factor must have been kept constant?

c) Which wavelength of light was (i) most (ii) least effective at bringing about an increase in gonadotrophic hormone production?

d) The light-sensitive pigment present in the hypothalamus is called **visual purple**. Suggest a reason for this based on the information given in the graph.

e) Which gland in the bird's body secretes gonadotrophic hormone?

f) State TWO effects that increased secretion of gonadotrophic hormone would have on a male bird.

6 Under natural conditions in northern latitudes, ferrets do not breed from September to March. However they come 'on heat' in April regardless of how cold the weather is.

a) Suggest which environmental factor triggers gonadal activity in ferrets.

b) If this is the case, are ferrets long day or short day breeders?

c) Predict whether a high or a low amount of melatonin would be secreted by a ferret's pineal gland during December.

d) Describe a possible effect of exposing ferrets to artificially prolonged photoperiods during mid-winter.

e) Why is daylength a more reliable indicator of time of year to an animal than temperature?

7 Some people who live in northern latitudes suffer deep depression in winter. This is thought to be linked to the high levels of melatonin known to be secreted by the pineal gland at this time.

During one mid-winter, a group of 34 such patients volunteered to be exposed to very bright light for several hours each day. Within a few days 30 of them found that their depression had lifted. It returned when the light treatment stopped.

Construct a hypothesis to account for these findings.

DATA INTERPRETATION

Healthy young specimens of four unrelated species of flowering plants (**A-D**) were subjected to a range of different light and dark treatments. The results are recorded in table 30.2 (where + = flowering, and − = no flowering).

	treatment					
	1	2	3	4	5	6
hours of light	15	14	13	12	11	10
hours of darkness	9	10	11	12	13	14
species A	−	−	+	+	+	+
species B	+	+	+	+	−	−
species C	−	−	−	+	+	+
species D	+	+	+	+	+	+

Table 30.2

One of these species was chosen for use in a further series of experiments. The details and results are displayed in figure 30.16.

a) (i) State the photoperiodic group to which each species of flowering plant **A-D** belongs. Where appropriate, give the critical number of hours of light or darkness that the species needs in order to flower. (4)

(ii) Which treatment in table 30.2 results in flower production by all four species of plant? (1)

b) Which of the four species was used for the experiments recorded in figure 30.16? Justify your choice of answer with reference to treatments **V** and **W** only. (1)

c) In terms of phytochrome, explain why no flowering occurred in plants subjected to treatment **X**, yet flowering did occur as a result of treatment **Y**. (2)

d) Predict the outcome of subjecting the plants to treatment **Z**. Explain your answer. (1)

e) Describe TWO treatments that could be employed in order to make plants of species **B** flower out of season. (1)

(10)

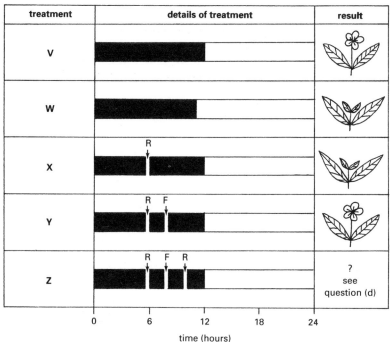

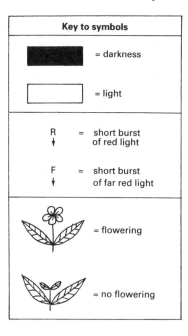

Figure 30.16

What you should know (CHAPTERS 29–30)

1 Nitrogen, phosphorus, potassium and magnesium are four **macro-elements** needed by plants for healthy growth.

2 A plant grown in a water culture experiment lacking one essential macro-element displays certain **deficiency symptoms**.

3 **Lead** inhibits the action of certain enzymes. It is **poisonous** to humans and in high concentrations may have a detrimental effect on the intelligence of children.

4 **Iron** is essential to humans for the formation of **haemoglobin**.

5 **Calcium** is essential to humans for the formation of healthy **bones** and **teeth**.

6 **Vitamin D** is needed by humans to promote the uptake of calcium ions by bone and prevent **rickets**.

7 **Drugs** such as thalidomide, alcohol and nicotine **harm** a developing human foetus.

8 A plant grown in **darkness** develops a weak elongated stem and small curled leaves and is said to be **etiolated**.

9 In order to flower, **long day** plants require a number of hours of light above a certain critical level. **Short day** plants need a critical number of hours of darkness.

10 The light sensitive pigment which governs the flowering response is called **phytochrome**. It exists in two forms, P_{660} and P_{730}.

11 **Florigen** is the name given to the hormone thought to initiate flowering.

12 To release florigen, **long day** plants need a high concentration of P_{730} and **short day** plants need a high concentration of P_{660}.

13 **Seasonal gonadal activity** in many birds and mammals is stimulated by the arrival of daily photoperiods of a certain length.

14 The **phenotype** of a developing organism is the product of the effects of **internal** factors (e.g. hormones) and **external** factors (e.g. light) acting directly or indirectly on the **genotype**.

Section 4 | Regulation in biological systems

31 Physiological homeostasis

Internal environment

A multicellular organism such as a human being consists of millions of cells bathed by extracellular fluid. This liquid (and the cells that it bathes) is called the **internal environment**.

Need for control

For the human body to function efficiently, the state of the internal environment must be maintained within tolerable limits.

For example, the **water concentration of blood** and the **concentration of cell chemicals** must be regulated so that they remain at a fairly constant level. Otherwise many physiological and biochemical functions of living cells such as nervous co-ordination and membrane permeability would be impaired. In addition, problems would be caused by osmotic imbalances.

Blood sugar level must be kept within a certain range of concentration to provide the energy needed by cells to perform energy-demanding jobs. These include synthesis of complex molecules (e.g. DNA, protein etc.), active uptake of ions, and muscle contraction.

In addition, the **temperature** of the human internal environment must be maintained at around 37°C to provide optimum conditions for the many enzyme-catalysed reactions of metabolism to proceed efficiently.

These and many other features of the internal environment are controlled by **homeostasis**.

Physiological homeostasis

This is the maintenance of the body's internal environment within certain tolerable limits despite changes in the body's external environment (or changes in the body's rate of activity).

Principle of negative feedback control

When some factor affecting the body's internal environment deviates from its normal optimum level (called the **norm** or **set point**), this change in the factor is detected by **receptors**. These send out nervous or hormonal messages which are received by **effectors**. The effectors then bring about certain responses which counteract the original deviation from the norm and return the system to its set point. This corrective mechanism is called **negative feedback control** (see figure 31.1).

Homeostasis in the human body

Osmoregulation

This is the means by which the body maintains its concentration of water, salts and ions at the correct level. Osmoregulation is under homeostatic control. The **kidneys** act as the effectors in the system by responding to hormonal messages and bringing about negative feedback control.

Control of water concentration of blood

If the water concentration of the blood decreases (due to increased sweating, lack of drinking water or consumption of salty food) then **osmoreceptors** in the **hypothalamus** are stimulated. These trigger the release of much **ADH (antidiuretic hormone)** by the **pituitary gland** into the bloodstream.

On arriving in the kidneys, ADH increases the

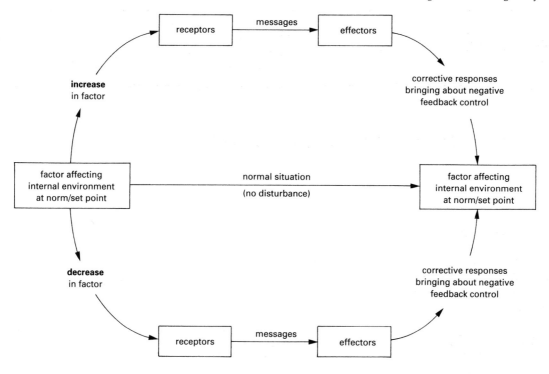

Figure 31.1   Principle of negative feedback control

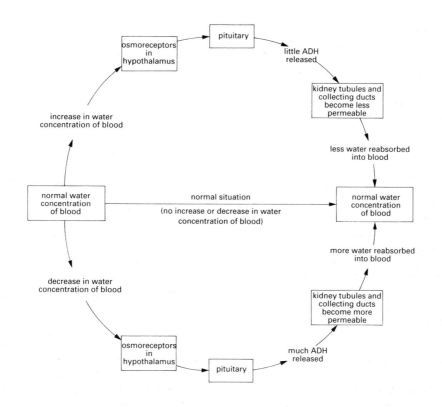

Figure 31.2   Homeostatic control of blood water concentration

permeability of the tubules and collecting ducts to water. As a result more water is reabsorbed into the bloodstream. This increases its water concentration until it is returned to normal.

If the water concentration of the blood becomes too high, less ADH is released and the kidney tubules and collecting ducts become less permeable to water. Less water is reabsorbed and the blood's water concentration soon drops to around its normal level.

This process is summarised in figure 31.2.

Control of blood sugar level

All living cells in the human body need a continuous supply of energy. Most of this energy is released by the oxidation of glucose (see page 21). Cells are therefore constantly using up the glucose present in the bloodstream (i.e. **blood sugar**).

However the body obtains supplies of glucose only on those occasions when food is eaten. To guarantee that a regular supply of glucose is available for use by cells, regardless of when and how often food is consumed, the body employs a homeostatic mechanism.

Liver as a storehouse.
Several hundred grams of glucose are stored as **glycogen** in the **liver**. Glucose can be added to or removed from this reservoir of stored carbohydrate depending on shifts of supply and demand.

Insulin and glucagon
A rise in blood sugar level to above its set point (e.g. following a meal) is detected by cells in regions of the **pancreas** called the **Islets of Langerhans**. These receptor cells produce **insulin**. This hormone is transported in the bloodstream to the liver where it activates an enzyme which catalyses the reaction:

glucose ⟶ glycogen

This brings the blood sugar concentration down to around its normal level.

If the blood sugar level drops below its set point (e.g between meals or during the night), different cells in the Islets of Langerhans detect this change and release **glucagon**. This second hormone is transported to the liver and activates a different enzyme which catalyses the reaction:

glycogen ⟶ glucose

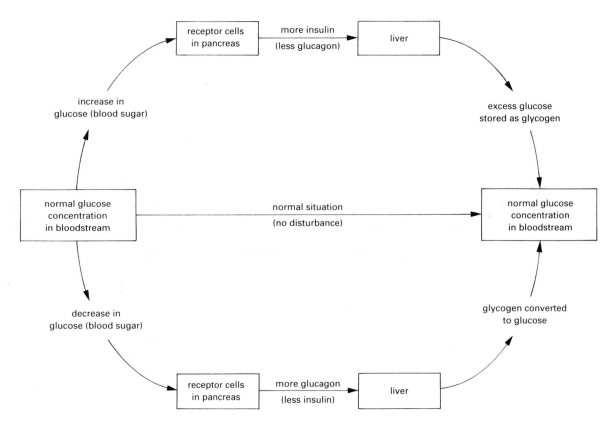

Figure 31.3 Homeostatic control of blood sugar level

The blood sugar concentration therefore rises to around its normal level. Figure 31.3 gives a summary of this homeostatic system.

Adrenaline

During an **emergency**, the body needs additional supplies of glucose to provide energy quickly for 'fight or flight'. On these occasions, the **adrenal glands** secrete an increased amount of the hormone **adrenaline** into the bloodstream.

Adrenaline overrides the normal homeostatic control of blood sugar level by inhibiting the secretion of insulin and promoting the breakdown of glycogen to glucose.

Once the crisis is over, secretion of adrenaline is reduced to a minimum and blood sugar level is returned to normal by the appropriate corrective mechanism involving insulin or glucagon.

Alternative diagram

Every factor under homeostatic control can be represented by the type of diagram shown in figure 31.1. However it must be kept in mind that when a factor deviates from its norm and is then returned to this set point by negative feedback control, it often overshoots the mark. This triggers the reverse set of corrective mechanisms.

To illustrate that a factor which is in a state of dynamic equilibrium is **constantly wavering on either side of its set point**, homeostasis is often represented as two interrelated circuits (see figure 31.4).

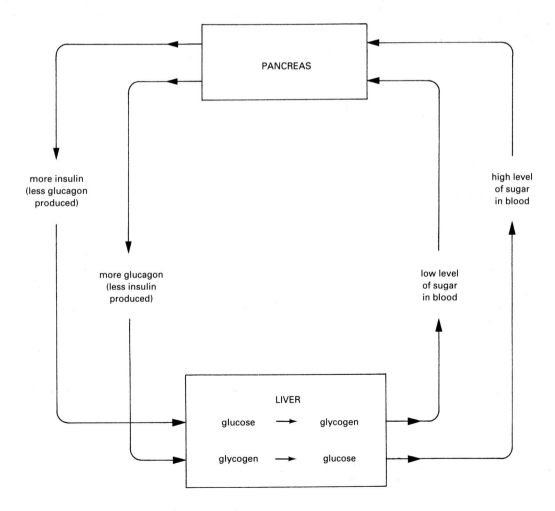

Figure 31.4   Alternative diagram of blood sugar control

Diabetes mellitus

Some people suffer a disorder known as *diabetes mellitus* because some (or all) of their insulin-secreting pancreas cells are non-functional. Since sufferers produce insufficient (or no) insulin, the concentration of glucose in their blood rises to 10–30 millimoles per litre compared with the normal concentration of 5 millimoles per litre.

The glomerular filtrate formed in the kidneys of a diabetic is so rich in glucose that much of it is not reabsorbed into the bloodstream. It is instead excreted in urine.

In the absence of insulin, cells are unable to use glucose efficiently and fat stores become depleted leading to loss in weight and wasting of tissues. Whereas *diabetes mellitus* used to be a fatal disorder, it is now successfully treated by regular injections of insulin and a controlled diet.

Control of body temperature

Ectotherm

An **ectotherm** is an animal which is *unable* to regulate its body temperature (by physiological means). All invertebrates, fish, amphibians and reptiles are ectotherms and their body temperature normally varies directly with that of the external environment.

Endotherm

An **endotherm** is an animal which is *able* to maintain its body temperature at a relatively constant level independent of the temperature of the external environment. All birds and mammals are endotherms. They have a **high metabolic rate** which generates much heat energy. Regulation of their body temperature is brought about by homeostatic control.

Role of hypothalamus

In an endotherm the hypothalamus acts as the **temperature-monitoring** centre. Its **thermoreceptors** detect changes in the temperature of the blood (which in turn reflect changes in the temperature of the body core). These thermoreceptors send nervous impulses to effectors which trigger corrective feedback mechanisms thereby returning the body temperature to its set point.

Role of skin

The **skin** plays a leading role in temperature regulation. It possesses heat and cold thermoreceptors which relay information about the external environment to the central thermoreceptors in the hypothalamus.

In response to nervous impulses sent from the hypothalamus, the skin acts as an effector. It corrects overheating of the body by employing the following mechanisms:

1 Increase in rate of sweating
Heat energy from the body is used to convert the water in sweat to water vapour and by this means brings about a lowering of body temperature.

2 Vasodilation
Arterioles leading to skin become dilated (see figure 31.5). This allows a large volume of blood to flow through the capillaries near the skin surface. From here, the blood is able to lose heat by radiation.

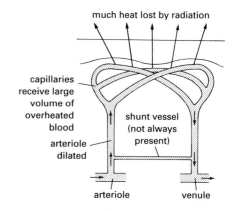

Figure 31.5 Vasodilation

The skin corrects overcooling of the body by employing the following mechanisms:

1 Decreased rate of sweating

2 Vasoconstriction
Arterioles leading to the skin become constricted (see figure 31.6). This allows only a small volume of blood to flow to the surface capillaries. Little heat is therefore lost by radiation.

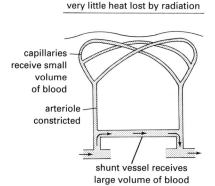

Figure 31.6 *Vasoconstriction*

3 Contraction of erector muscles

This process (see figure 31.7) results in hairs being raised (or feathers fluffed out). A wide layer of air (which is a poor conductor of heat) is trapped between the animal's body and the external environment. This layer of **insulation** reduces heat loss.

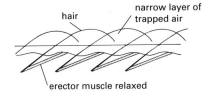

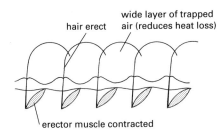

Figure 31.7 *Action of hair erector muscles*

The homeostatic control of body temperature is summarised in figure 31.8. It includes further corrective mechanisms such as **shivering** and **changes in metabolic rate**.

Investigating response to sudden heat loss (using a thermistor)

A **thermistor** is a device which responds to tiny changes in temperature. In this investigation the thermistor is taped between two fingers of one hand as shown in figure 31.9 and the initial temperature of the skin is recorded from the digital meter.

The other hand is plunged into a container of icy water to cause a sudden heat loss.

Temperature readings are taken every 30 seconds for five minutes; the skin temperature of the hand attached to the thermistor is found to drop by 1–2°C. (If a second thermistor is positioned in the armpit during the experiment, the temperature of the body core is found to remain constant.)

It is therefore concluded that when heat is lost from one extremity (e.g. hand in icy water), a **compensatory reduction in temperature** occurs in the other extremity (but not in the temperature of the body core).

This reduction in temperature is brought about by the following homeostatic mechanism: thermoreceptors in the skin in icy water send nervous impulses to the hypothalamus which in turn sends impulses to the other hand causing **vasoconstriction** which reduces heat loss.

This response by the body's extremities helps to conserve heat when the body is exposed to extremes of temperature. The temperature of the body's extremities is therefore found to fluctuate more than that of the body core.

Advantage of homeostasis

The evolution of systems that respond to homeostatic control is of **survival value** because these maintain the body's internal environment at a relatively steady optimum state. This allows the body to function efficiently despite wide fluctuations in the external environment (many of which would be unfavourable).

Extreme conditions

Homeostasis only works *within certain limits*. If a person is exposed for a *prolonged* period to extremely adverse conditions in the external environment (e.g. freezing temperatures, total lack of drinking water etc.), the body responds by exerting negative feedback control as described above. However a homeostatic system eventually breaks down when its corrective mechanisms can no longer return the body to its steady state. In extreme cases, death results.

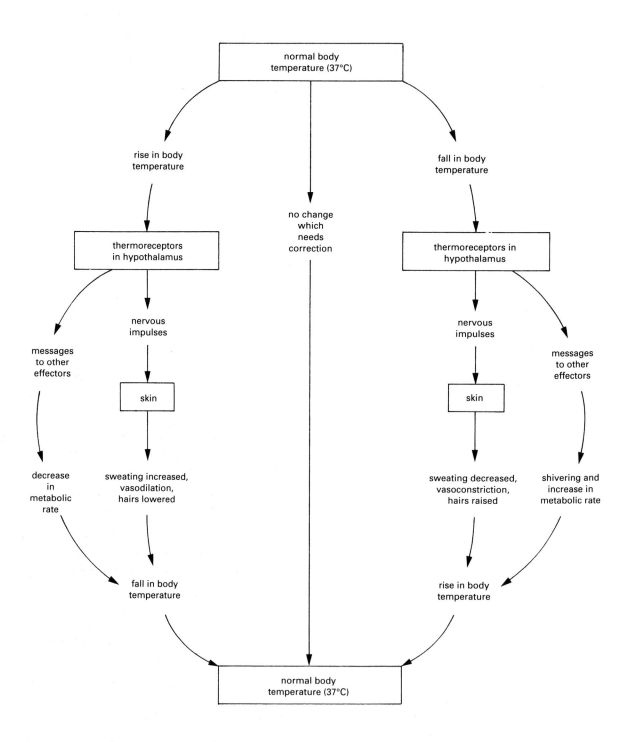

Figure 31.8   Homeostatic control of body temperature

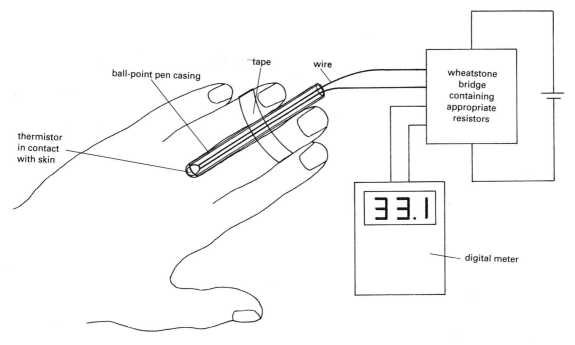

Figure 31.9   Use of thermistor

QUESTIONS

1 **a**) Define physiological homeostasis.
b) Outline the principle of negative feedback control.
c) Why is such control of advantage to an organism?
2 Figure 31.10 shows a simplified version of the homeostatic control of blood water concentration.

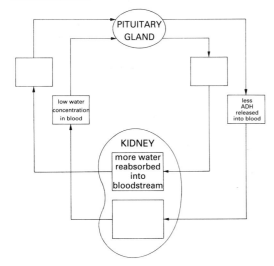

Figure 31.10

a) Redraw the diagram and complete the blanks.
b) (i) Name the hormone represented by the letters ADH and state the effect that it has on kidney tubules.
(ii) What relationship exists between amount of ADH in the bloodstream and volume and concentration of urine produced by the kidneys? Explain your answer.
c) Figure 31.10 omits reference to the osmoreceptors that detect changes in the blood's water concentration. Where in the body are they located?
d) In addition to their role in the above control system, the osmoreceptors send nervous messages to the cerebrum producing the sensation of thirst. Explain how this could act as a successful corrective mechanism.
3 Figure 31.11 shows the effect of consuming 50 g of glucose (after a period of fasting) on the concentrations of fatty acids, glucose and insulin in the bloodstream.
a) (i) During which period of time was the person's blood sugar concentration at a steady level?
(ii) By what means is this steady level maintained?
b) (i) At what time was the glucose consumed?
(ii) What initial effect did the intake of glucose have on blood sugar level and amount of insulin in the blood?

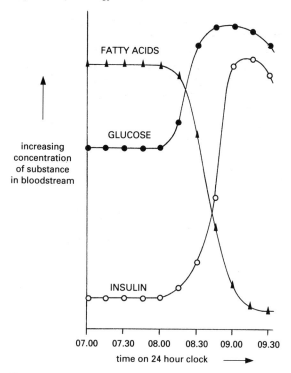

FATTY ACIDS

GLUCOSE

increasing
concentration
of substance
in bloodstream

INSULIN

07.00   07.30   08.00   08.30   09.00   09.30

time on 24 hour clock ⟶

Figure 31.11

(iii) Why was there a short time lag between these two effects?

c) What evidence is there from the graph that insulin suppressed the breakdown of stored fat?

4 Hormones are chemical messengers released directly into the bloodstream by endocrine glands.

hormone	endocrine gland from which hormone originates	letter(s) indicating effect(s) of hormone
adrenaline		
insulin		
ADH		
glucagon		

Table 31.1

a) Copy table 31.1 and complete the central column.

b) Complete the right hand column by using a selection of one or more answers from the following list:

A promotes conversion of excess glucose to glycogen;

B increases permeability of kidney collecting ducts;

C promotes conversion of glycogen to glucose;

D decreases blood sugar level;

E prepares the body to cope with an emergency.

5 The data shown in table 31.2 was obtained during an investigation into the functions of the human kidney.

	concentration of substance present in glomerular filtrate (g/100 cm³)	concentration of substance present in urine (g/100 cm³)
sodium ions	0.3	0.6
glucose	0.1	0.0
urea	0.03	2.10

Table 31.2

a) With reference to the homeostatic control of osmoregulation, explain why the concentration of urea is higher in urine than in the glomerular filtrate.

b) By how many times is the concentration of (i) sodium ions, (ii) urea, more concentrated in urine than in the glomerular filtrate?

c) Suggest why the concentration factor affecting urea is higher than that affecting sodium ions.

d) State ONE way in which the data in table 31.2 would differ if it referred to a newly diagnosed sufferer of *diabetes mellitus* prior to treatment.

(ii) What treatment would be given?

6 With reference to homeostasis, explain why a person suffering from prolonged overcooling may eventually die of hypothermia.

DATA INTERPRETATION

Sodium accounts for about 90% of the positive ions in the human body's extracellular fluid. Figure 31.12 shows how the hormone aldosterone helps to maintain the sodium concentration of the extracellular fluid.

a) Give an example of an extracellular fluid present in the human body.   (1)

b) Identify the (i) receptor, (ii) effector, in the example of homeostasis shown in the diagram.   (1)

c) What type of message passes from the receptor to the effector in this case?   (1)

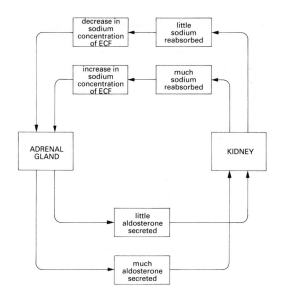

Figure 31.12

d) Name the corrective mechanism which restores the sodium concentration to its set point when salt intake is too high. (1)

e) Redraw the diagram in the form illustrated in figure 31.1 on page 175. (2)

f) (i) A reduction in the volume of the body's extracellular fluid, caused by diarrhoea or a haemorrhage, is followed by an increase in aldosterone production. Suggest why. (1)

(ii) This increase in aldosterone production brings about a decrease in the blood's water concentration which, in turn, triggers increased production of a second hormone. Identify the latter. (1)

g) Predict with reasons the ultimate fate of a person fed a sodium-free diet for a prolonged period. (2)

(10)

32 Population dynamics – part 1

A **population** is a group of individuals of the same species which makes up part of an ecosystem. The number of individuals present per unit area (or unit volume) of a habitat is called the **population density**.

The **birth rate** of a population is a measure of the number of new individuals produced during a certain interval of time. The **death rate** is a measure of the number of individuals that died during the same interval of time.

Population dynamics is the study of population changes (growth, maintenance and decline) and the factors which cause these changes.

Stability

When a population colonizes a new environment, it grows in number until it reaches a certain size which the available environmental resources can just maintain. This limit is called the **carrying capacity** of the environment.

The population is now in a state of **dynamic equilibrium** and remains relatively stable

despite short-term oscillations in number from generation to generation. Birth rate now equals death rate and the population neither increases nor decreases in size.

Sheep

Figure 32.1 shows the growth curve of the population of sheep following their introduction to the Australian island of Tasmania in the nineteenth century. Although the numbers varied from one census to another, the population remained fairly stable once it had reached the carrying capacity of the environment in the 1850s.

Birds

More recent long-term studies have been made of bird populations such as tawny owls in an Oxford wood and herons in the Thames Valley. In each case the results reveal a picture of a stable population which, in the absence of human intervention, fluctuates only slightly over long periods of time.

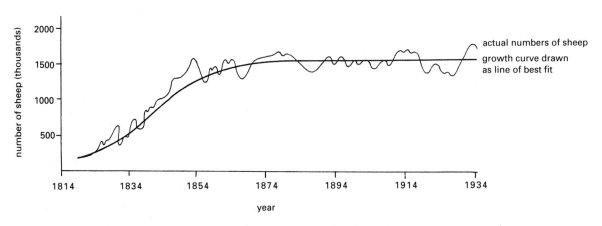

Figure 32.1 Population growth and stability

Factors influencing population change

Environmental resistance

Each species has an enormous reproductive potential (see table 32.1). However under natural conditions in an ecosystem, a population is prevented from increasing in size indefinitely by environmental resistance. This consists of factors such as: availability of food, water, oxygen, light, space and shelter; predation; disease; and climate. These factors which affect the size of a population fall into the following two categories.

animal	average number of offspring per year
fox	5
red grouse	8
rabbit	24
mouse	30
trout	800
cod	4 000 000
oyster	16 000 000

Table 32.1   Reproductive potential

Density-independent factors

A **density-independent** factor is one which affects the growth of a population regardless of the population's density. A forest fire, for example, will wipe out most of a population whether it is densely packed or sparsely distributed in an ecosystem.

Similarly extremes of weather such as storms, floods, drought and spells of unusually high or low temperature all act as density-independent factors. They tend to cause a sudden drastic reduction in population number. However, given sufficient time, the population normally makes a full recovery.

Density-dependent factors

A **density-dependent** factor is one which only affects the population once it has grown to a certain size (and density). In the absence of environmental resistance, a small population continues to grow until it reaches a certain density. Its growth rate is then affected by one (or more) of the following density-dependent factors each of which has a regulatory effect.

Shortage of food
When the number of animals present in a population outstrips the **available food supply**, competition occurs between individuals (see page 220) and some starve to death.

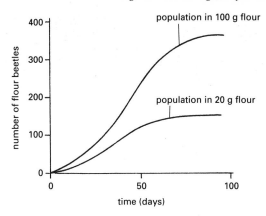

Figure 32.2   Effect of food shortage on population growth

Experiments using flour beetles show that when the amount of food per beetle drops to below a certain level, frequency of mating declines. Egg production decreases and cannibalism sets in. As a result population growth is affected (see figure 32.2).

Toxic wastes
At high population densities, toxic wastes made by the members build up and create poisonous conditions which prevent further growth of the population. This is often the factor which limits the growth of a population of micro-organisms cultured in food-rich conditions at optimum temperature.

Parasitism and disease
A **parasite** is an organism which benefits at the expense of another organism (the **host**) by feeding on it. The denser the population of the host, the greater the chance of the parasite obtaining food and producing offspring which will in turn find new hosts.

Compared with a sparse population, a dense host population is more prone to attack and damage by parasites. Some parasites cause disease directly whereas others transmit disease-causing micro-organisms from one host to another. Either way, the subsequent disease reduces the host population in number.

A dense population of a domesticated organism (e.g. a crop of cereal plants of the same genotype) grown as a vast monoculture is especially vulnerable to parasitic attack.

Predation
A dense population of **prey** organisms is more prone to attack by their **predators**. Figure 32.3 summarises the results from a survey carried out on the nests of the great tit. Weasels smash open the bird's eggs and eat the chicks.

From the graph it can be seen that a higher

185

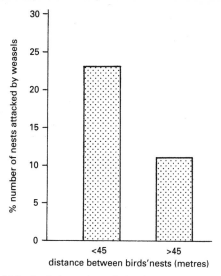

Figure 32.3 *Predation and population density*

number of nests are attacked when they are close together than when they are spaced out. This shows that predation by weasels is a density-dependent factor.

Predator-prey interactions

A delicate balance exists between populations of predators and their prey. An increase in number of prey (perhaps due to climatic conditions favouring growth of their plant food) leads to an increase in predation.

As the size of the prey population decreases, competition between predators for the remaining prey becomes more and more intense until eventually the number of predators drops. This in turn allows the prey population to build up again which leads to a corresponding increase in the predator population and so on.

This self-regulating series of **interdependent**

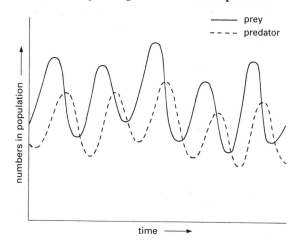

Figure 32.4 *Predator–prey interactions*

fluctuations is summarised in figure 32.4. The predator curve mirrors that of the prey but lags behind it since time is required for each change in the sequence of events to take effect.

Two examples of animals showing this relationship are *Hydra* which feeds on waterfleas (see figure 32.5) and lynx which preys upon the snowshoe hare (see data interpretation on page 190).

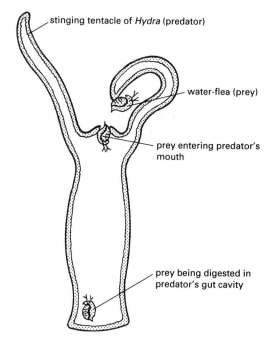

Figure 32.5 *Predator and prey*

In a natural environment, the predator-prey relationship is not always as straightforward as suggested by figure 32.4. A prey organism is often hunted by several predators. In addition, predators will turn to alternative prey if their favourite food is in short supply.

Investigating the effect of soil type on population density of springtails

Springtails are tiny soil insects (see figure 32.6). Those which live in the air spaces in top soil possess a springing organ which enables them to jump an enormous distance relative to their body size. Population density of springtails can be measured as number of animals/cm³ of soil.

In this investigation the two soil types are 'cultivated' soil in an empty flower bed and nearby 'uncultivated' soil under dense bushes.

Samples are taken at random from both soils using a corkscrew soil **auger**. The auger is

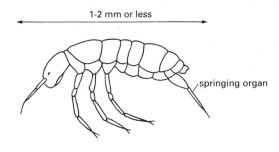

Figure 32.6   Springtail

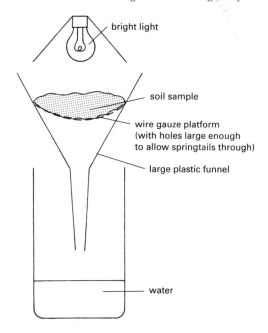

Figure 32.7   Tullgren funnel

screwed in a clockwise direction into the soil to a depth of 100 mm and then pulled up carefully. The soil clinging to it is transferred to a plastic measuring cylinder. This container is gently tapped to compact the sample until it resembles the soil in the sample site.

The volume of each soil sample is recorded and the soil kept for further use in a plastic bag. The moisture content of the soil at each sample site is recorded using a **moisture meter** (with a scale of 1 to 8 where 1 = dry and 8 = wet).

Back in the laboratory, each soil sample is transferred to a **Tullgren funnel** (see figure 32.7). Springtails (and other tiny animals) present in the sample move downwards away from the light source and fall through the holes in the gauze platform into the water.

After a standard length of time (e.g. 24 hours), the number of springtails extracted from each sample is recorded and the animals returned to their natural habitat. Table 32.2 gives a typical set of results.

Discussion of results

In this investigation, the springtails were found to be more densely populated in the moister soil under the bushes. Since these animals lose water through their permeable skins and are killed by severe drought (acting as a density-independent factor), it is possible that their population density is affected by the **moisture content** of the soil.

However it must be remembered that some other difference between the two soils could also affect the population density of the animals.

soil type	sample site	moisture meter reading	number of springtails	volume of soil sample (cm³)	population density (springtails/cm³)	average population density (springtails/cm³)
'cultivated' soil in empty flower bed	1	4	2	45	0.04	0.03
	2	3	3	50	0.06	
	3	4	1	50	0.02	
	4	5	0	45	0.00	
	5	4	3	55	0.05	
'uncultivated' soil under bushes	6	6	10	50	0.20	0.19
	7	7	8	45	0.18	
	8	7	11	45	0.24	
	9	6	9	50	0.18	
	10	6	8	50	0.16	

Table 32.1   Springtail population density results

Springtails feed on dead **organic matter** which is less plentiful in the flower bed (and may therefore be acting as a density-dependent factor).

In addition, many springtails are probably killed by the **repeated disturbance** of the flower bed during cultivation (acting as a density-independent factor). To find out exactly which factor (or combination of factors) determines the population density of springtails would require a complex series of experiments each involving one altered variable at a time.

Homeostatic control

Regulation is the tendency of a population to decrease in size when it rises above a certain level and increase in size when it falls below that level.

In a natural ecosystem, such control is brought about by one or more density-dependent factors acting on the population. This is a further example of **homeostasis** (negative feedback control) as shown in figure 32.8. It is the means by which a population is maintained at a relatively stable equilibrium, enabling it to make maximum use of available resources such as energy.

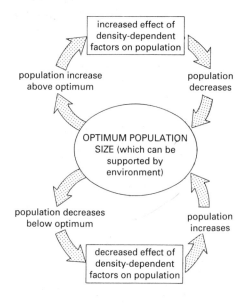

Figure 32.8 Homeostatic control of population size

QUESTIONS

1 The graphs in figure 32.9 show the result of population censuses carried out over a period of 20 years.
Graph **A** refers to a species of bird living in a wood.
Graph **B** refers to a species of flowering plant living on a moist meadow.

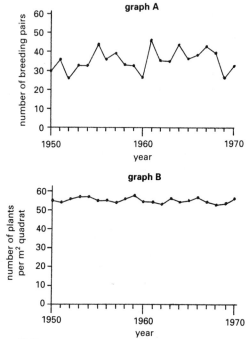

Figure 32.9

a) State the population density of the plant species in 1960.
b) Estimate the environment's carrying capacity of the bird species.
c) Suggest why the censuses were continued for such a long period of time.
d) Rewrite the following generalisation made from the graphs by completing the blanks using the words given in brackets.

Once the _____ size of a species of _____ or animal has reached the _____ capacity of the _____, it remains relatively _____ despite short-term _____ in number.

(carrying, environment, oscillations, plant, population, stable)

2 a) Define the terms (i) density-dependent factor, (ii) density-independent factor.
b) Imagine a swarm of one million locusts. This

population could be decreased in number by increased predation, a thunderstorm, intense drought or shortage of food. For each of these factors, state whether it operates in a density-dependent or density-independent manner.

3 The graph in figure 32.10 shows the fluctuations in a population of phytoplankton (microscopic algae) in a Scottish loch.

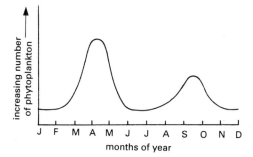

Figure 32.10

a) Name TWO density-independent factors that could have been responsible for the sudden rise in numbers of the plankton in spring. Explain your answers.

b) These plants are eaten by water fleas. Make a copy of the graph and then draw in a second curve to show the likely fluctuations in water flea population over the course of one year.

4 a) Name THREE factors which could exert a regulatory effect on a population of herbivorous animals thereby preventing it from increasing indefinitely in size.

b) What general name is given to this process of negative feedback control?

5 The graph in figure 32.11 shows the results from a study of the populations of two organisms, a predator and its well fed prey, over a period of several weeks.

a) (i) How many days did the predator population take to reach its maximum size?
(ii) By how many individuals did it increase in number during this period of time?

b) By how many times did the prey outnumber the predators at week 4?

c) Account for (i) the decrease in number of predators between weeks 7 and 8;
(ii) the increase in number of prey between weeks 8 and 9.

6 In 1954–55, 90% of Britain's rabbit population died of a viral disease called *myxomatosis*. The virus was passed from animal to animal in underground burrows by the rabbit flea. The survivors were those rabbits that lived a relatively solitary existence above ground and rarely caught one another's fleas.

a) Is *myxomatosis* a density-dependent or density-independent factor governing the size of a rabbit population? Explain your answer.

b) (i) Construct a food web to include the following organisms: fox, grass, rabbit, sheep, tree seedling.
(ii) *Myxomatosis* disturbed this food web for many years. Suggest ONE advantage and ONE disadvantage to the farmer that resulted during this time.

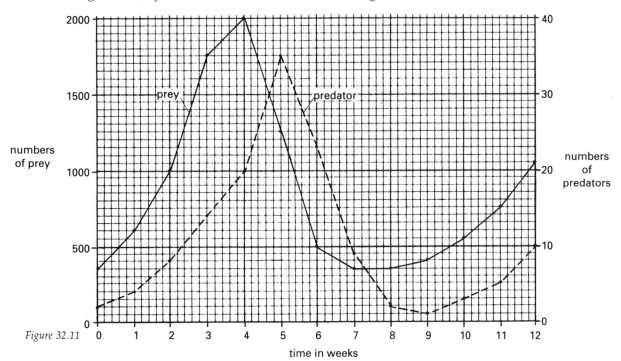

Figure 32.11

EXPERIMENTAL DESIGN QUESTION

The fruit bushes in a garden were sprayed once with a 1% solution of a new pesticide called NABBEM.

Observations: A few days later the soil beneath the bushes was found to contain very few springtails compared with similar soil under fruit bushes which had not been sprayed.

Hypothesis: Most of the springtails under the bushes sprayed with NABBEM were killed by pesticide which had dripped down onto the soil.

Problem: Design an experimental procedure to test the validity of the above hypothesis.

If the results of your investigation were found to support the hypothesis, what advice would you give the gardener regarding the use of NABBEM in the future?

DATA INTERPRETATION

The following information refers to two separate examples of predation. The graph in figure 32.12 shows the fluctuations in population numbers of the snowshoe hare and the lynx (the former making up 80–90% of the latter's diet).

The diagram in figure 32.13 shows the results of removing starfish (the dominant predator) from a marine food web.

a) (i) From figure 32.12, identify which line represents the predator and which the prey. (1)

(ii) Give TWO reasons to justify your choice. (2)

b) (i) Which of the following periods (in years) is closest to the average length of a complete population cycle of the prey animal: 5; 10; 25; 50? (1)

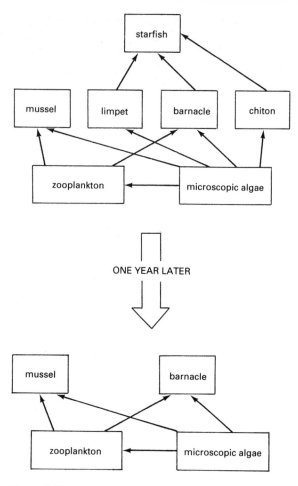

Figure 32.13

(ii) Account for the regular fluctuations in population number of predator and prey. (1)

(iii) It is now known that the regular cycles in prey numbers still occur in the absence of predators. The plants eaten by the hares respond to heavy grazing by producing unpalatable shoots rich in toxins for 2–3 years. Suggest therefore why some scientists prefer to describe the predator cycles as *tracking* rather than *causing* the prey cycles. (1)

c) With reference to figure 32.13:
(i) Name TWO prey animals that disappeared from the food web after one year. (1)
(ii) Suggest why they disappeared. (1)
(iii) Why, in this case, is the presence of the predator *beneficial* to these prey? (1)

d) In the two different examples of predation above, which predator is a 'generalist' and which is a 'specialist'? (1)

(10)

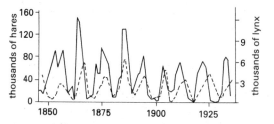

Figure 32.12

33 Population dynamics – part 2

Monitoring populations.

Many species are kept under close surveillance by scientists in order to obtain information about their population numbers and the factors which affect them. This applies especially to:

- species which provide humans with valuable sources of **food**;
- species which are **endangered** and in need of protection and conservation in case they become extinct;
- species which act as **indicators** of the state of the environment's health.

Food species

If the death rate of a species exceeds its birth rate then the size of the population decreases. In the long term, this leads to extinction of the species.

The **maximum sustainable yield (MSY)** of a population is a measure of the rate at which individuals can be removed without affecting the population's future productivity.

Fish

Populations of edible fish species such as **herring** are monitored to estimate their MSY since it is essential that fish are not removed at a rate which exceeds their maximum rate of reproduction.

Red deer

Scotland has a large population of **red deer** living in the wild. Although the very young members of the population are vulnerable to attack by foxes and golden eagles, wild predators mostly make no impression on the adult numbers. In an attempt to prevent uncontrolled population growth occurring amongst red deer, humans kill several thousand animals every year for their meat (venison). This planned harvesting of a wildlife species is called **culling**.

In the early 1970s, the number of Scottish red deer stood at around 200 000 which was regarded by experts as an optimum level. However by 1990, the population had risen to an estimated 300 000 despite a cull of 3000 animals every year. Many experts therefore believe that the annual cull should now be increased to keep the population stable and boost the supply of a valuable food resource.

Endangered species

The monitoring of populations of wild plants and animals enables humans to recognise and protect rare species for their **aesthetic value** and **genetic diversity** (see page 100).

Wild plants

In Britain, 21 native species of **wild plants** are now known to be very rare and in need of rigorous protection. Two examples are shown in figure 33.1.

To prevent such endangered species from becoming extinct, the areas that they live in are often designated as **nature reserves**. In addition, the Conservation of Wild Creatures and Wild Plants Act (1975) makes it illegal to remove *any part* of these plants since picking even just one flower or leaf could reduce the plant's chance of survival.

Black rhinoceros

Figure 33.2 shows the dramatic decline in number of the **black rhinoceros** of Kenya in recent years. At last steps are being taken to save this animal which has been over-hunted and is presently threatened with extinction.

Whale

Although estimates of **whale** populations tend to be subject to a large margin of error, it is known that their numbers have dropped dramatically since the turn of the century (also see table 33.1).

This downward spiral will only be reversed in the future if the current **moratorium** on

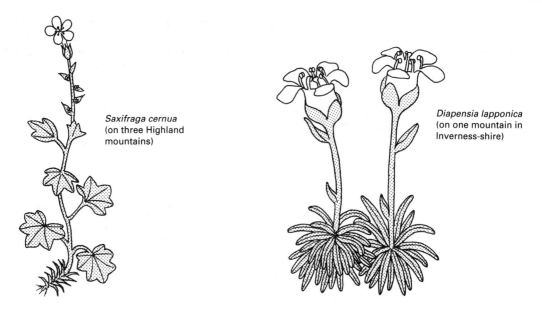

Figure 33.1   Rare wild plants

species of whale	estimated number before commercial whaling	estimated number today	percentage remaining
Blue	200 000	2 000	1
Humpback	100 000	9 000	9
Fin	450 000	70 000	15.6
Sei	200 000	28 000	14
Right	50 000	3 000	6

Table 33.1   Decline in whale numbers

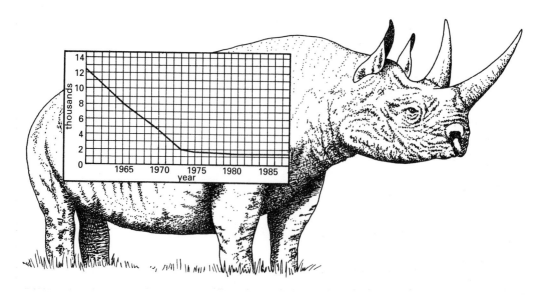

Figure 33.2   Population decline of black rhinoceros

commercial whaling is respected by all countries of the world.

More is at stake than the survival of the world's largest mammal. Like the panda and the elephant, the whale has become an emblem of conservation. Experts fear that if these 'high profile' species cannot be saved then there is little hope of preventing a multitude of other species of 'lower profile' (but of equal importance) from slipping into extinction (see also page 111).

Indicator species

Certain wildlife species often serve as **indicators** of the state of the environment, for example by their abundance or scarcity.

Fresh water invertebrates
The presence of large populations of **mayfly** and **stonefly** in a river ecosystem shows that the water is clean and rich in dissolved oxygen. On the other hand, an abundance of **rat-tailed maggots** and **sludgeworms** in the water indicates that it has a low oxygen content and is badly polluted with organic waste.

Lichens
Lichens are simple plants which grow on the trunks and branches of trees. Three examples are shown in figure 33.3.

Since lichens vary in their sensitivity to **sulphur dioxide** (SO_2), they indicate the level of

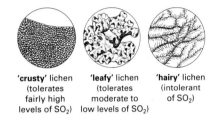

'**crusty**' lichen (tolerates fairly high levels of SO_2) '**leafy**' lichen (tolerates moderate to low levels of SO_2) '**hairy**' lichen (intolerant of SO_2)

Figure 33.3   Lichens

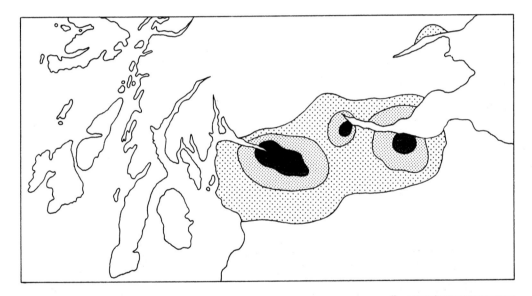

Key	abundance of lichen types	level of atmospheric pollution indicated
■ (black)	no lichens or few crusty only	high
(dark stipple)	many crusty, few leafy	moderate
(light stipple)	some crusty, many leafy	low
(white)	some crusty, many leafy, some hairy	zero

Figure 33.4   Air pollution map based on lichen survey

atmospheric pollution by this harmful gas. Using the data obtained from a lichen survey, an air pollution map of an area can be constructed as shown in figure 33.4.

Birds of prey

Predatory birds at the top of their food chain (e.g. peregrine falcons) are the first to suffer the consequences of environmental degradation by over-use of pesticide sprays on crops. High concentrations of chemical residues accumulate in the birds' tissues and indicate the state of the local environment's health.

Phytoplankton

These microscopic plants inhabit the world's oceans and act as the first link in marine food chains. Since they absorb vast amounts of carbon dioxide during photosynthesis, it is possible that they are helping to protect the world against the greenhouse effect (i.e. overheating as a result of rising levels of atmospheric carbon dioxide from the combustion of fossil fuels).

Scanners fitted to weather satellites reveal that enormous numbers of phytoplankton occur in the oceans at certain times of the year. For example, explosive blooms appear in the North Atlantic in spring when warmth and sunlight coincide with rich supplies of nutrients stirred up from the deep by winter storms.

It is important to monitor the populations of these tiny producers because they indicate the state of the ocean ecosystem. The future survival of the earth may even depend on their activities if tropical forests continue to be cleared at the present rate.

Case study of a population

In recent years, millions of tons of poisonous wastes have been pumped or dumped into the North Sea. These have included sewage sludge, fertilisers, incinerator fall out, heavy metals, organochlorines, dioxins, radioactive waste and oil.

In 1988 populations of the **common seal** were found to be suffering from a **virus** (unique to seals but similar to canine distemper) which attacked their immune system. This led to an epidemic which wiped out an estimated 60% of the seal population.

Many scientists are cautious about linking this disease with pollution since, strictly speaking, the seals died of a natural cause. In 1989 blood samples were taken from common seals and about half of the animals tested were found to

possess antibodies to the disease. This meant that the others (mostly young pups) were susceptible to the virus. This proved to be true later that year when the disease struck again, especially amongst young animals.

Poisoned food chain

Organochlorines are toxic chemicals used as pesticides. Since they are **non-biodegradable**, they persist and accumulate in food chains with the final consumer being the most severely affected. This is shown in figure 33.5.

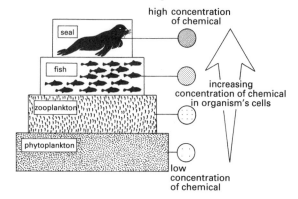

Figure 33.5   Accumulation of chemical in food chain

Although certain organochlorines such as DDT have now been banned from use by most countries, they still persist in almost all ecosystems. Many of the North Sea common seals that died of the virus were found to contain high concentrations of organochlorines and other pollutants in their tissues and blubber.

Some scientists are of the opinion therefore that an indirect link exists between the viral disease and pollution. They argue that animals living in an ecosystem under such constant stress by pollution must be more susceptible to disease. Some experts suggest that the pollutants may, in some way, impair the seals' immune system.

The future

In 1989/90, Britain agreed to:

- ban the incineration of toxic waste at sea by the end of 1992;
- aim to phase out the dumping of industrial waste in the North Sea by the end of 1992;
- stop dumping sewage sludge in the North Sea by 1998;
- clean up raw sewage discharges into the sea by British coastal towns;
- help to draw up international plans to protect seals (and other wildlife) in the North Sea.

QUESTIONS

1 Scientists *monitor* populations of certain species.
   a) What does this mean?
   b) Give THREE reasons why it is necessary.
2 Suggest why it is important that every effort is made during commercial fishing to net only the larger members of a shoal and leave the smaller ones in the sea.
3 **a)** What is meant by the expression *'moratorium on commercial whaling'* (page 191)?
   b) Some countries in the world still refuse to heed the moratorium. Why is it essential that they are persuaded to do so in the future?
4 The graph in figure 33.6 shows the results of an analysis of the breast muscles of several species of birds for an organochlorine pesticide.

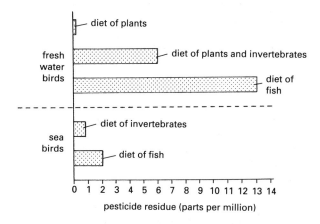

Figure 33.6

   a) What is meant by the term *'indicator species'*?
   b) (i) From the graph, state the general relationship that exists between a bird's diet and the concentration of pesticide residue in its muscles tissues.
   (ii) Account for this relationship.
   c) Which of the environments was less severely affected by the pesticide, as indicated by this data? Suggest why.
5 Read the following case study and answer the questions that follow:

   Populations of African elephant decreased in number from a total of about 1.2 million in 1979 to about 600 000 in 1989. For this reason many industrialised nations have declared a moratorium on ivory imports. African countries such as Kenya, whose elephant herds have been depleted in recent years by uncontrolled poaching, have led the call for this worldwide ban. They argue that as long as any legal trade in ivory exists, poachers and smugglers will find a way to beat the system.

   However in Zimbabwe, careful management has increased the elephant population from 30 000 in 1962 to around 51 000 in 1989. So successful has this conservation scheme been, that Zimbabwe now has to carry out an annual cull of several thousand elephants to keep herds at an optimum level and prevent the destruction of trees and grassland.

   Government ministers in Zimbabwe claim that their country is being punished for corruption and inefficiency in other African States. They therefore oppose the ban insisting that it would bring an end to Zimbabwe's prosperous elephant hide tanning business and deprive impoverished communities of jobs and supplies of free elephant meat. As a result, the incentive to protect elephant herds would be lost.

   a) Why do many countries support the international ban on ivory products?
   b) What is the well intentioned idea behind this moratorium?
   c) Explain why the ban seems unfair to Zimbabwe.
   d) Calculate the percentage increase in Zimbabwe's elephant population that occurred between 1962 and 1989.
   e) What is meant by the term 'culling'?
   f) In Zimbabwe what benefits are gained by (i) the local people, (ii) the environment, from the culling of elephant herds?
   g) What solution would you offer to the conflict of interests described in the above case study?

DATA INTERPRETATION

The following graphs in figure 33.7 show some of the effects of adding excessive amounts of untreated sewage to a river.
a) Broadly speaking, what relationship exists between concentration of dissolved oxygen and numbers of bacteria present in the river? Explain why.   (1)

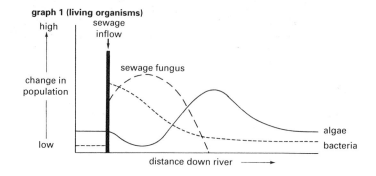

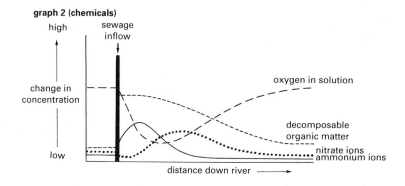

Figure 33.7

b) (i) What evidence is there from **graph 1** that sewage fungus is not a native member of the river community?
(ii) Account for the initial increase in population size of sewage fungus.
(iii) Name a factor that could have been responsible for the subsequent decrease in sewage fungus.
(iv) State whether the factor you gave as your answer to (iii) would operate in a density-dependent or density-independent manner. (4)
c) (i) What factor could account for the initial decrease in number of algae following the sewage inflow?

(ii) Is this factor density-dependent or density-independent?
Explain your answer. (2)
d) With reference to the nitrogen cycle, explain why the maximum concentration of nitrate ions occurs in the water *after* the maximum concentration of ammonium ions. (1)
e) Suggest which factor was responsible for the algal population boom. Why should this be so? (1)
f) Name a natural outside factor that could hasten the river's return to a normal healthy state. (1)

(10)

34 Population dynamics – part 3

Plant succession

Simplified example

Imagine a bare infertile field left abandoned to nature. Although many types of seeds land on it from neighbouring ecosytems, at first the only successful plants are annual (and a few perennial) **weeds**. If, for example, acorns from a neighbouring oak forest arrive, the seedlings fail. This is because the soil is poor and too exposed (hot and dry in summer; early to freeze in winter).

After a short number of years, the pioneer community of weeds is gradually replaced by sturdy **grasses** and taller **shrubby perennials** which become dominant by choking out the smaller plants and depriving them of light.

Amongst this community, **oak seedlings** are eventually able to survive once the soil becomes richer and moister and the shrubs become bushy enough to provide shelter. After many years, the oaks develop into trees whose dense canopy of leaves prevent much light reaching the ground.

This causes the majority of grasses and shrubs to die out leaving only shade-tolerant plants (e.g. bluebell) on the woodland floor.

Thus many years after the original field was abandoned, an **oak forest** stands in its place. The oak forest is called the **climax community**. This sequence of events is summarised in figure 34.1. Such a change involving a regular progression from a pioneer community of plants to a climax community is called **succession**.

Primary and secondary succession

Primary succession occurs during the colonisation of a **barren area** which has not previously been inhabited e.g. bare rock. In this case the pioneer community consists of lichens which are able to withstand drying out. They also make acids which break down the rock. Mosses are the next stage in the succession and, as the rock disintegration process continues and dead plants accumulate, a layer of proper soil gradually develops.

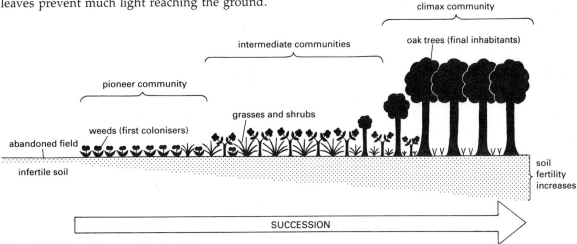

Figure 34.1   Succession on an abandoned field

Eventually the mosses are succeeded by small flowering plants and then shrubs and trees as before. Such primary succession takes a considerable length of time (e.g. a thousand years).

Secondary succession occurs during the colonisation of an area which has *previously been occupied* by a well developed community but has become barren. For example, a forest cleared for agricultural use and later abandoned by the farmer or a forest destroyed by fire.

Secondary succession takes a shorter time (e.g. one hundred years) because soil containing nutrients is already present and the conditions are normally more favourable to colonisation.

Cause of succession

The main driving force behind succession is the effect that the plants have on the habitat. Succession occurs because each community acts on and *changes* the habitat. After a period of relative short-term stability, a community ends up making the habitat less favourable for itself and more favourable for a different community which therefore succeeds it.

For example, during the early stage of the succession shown in figure 34.1, the members of the pioneer community add humus (dead organic matter) to the soil. This improves its texture and mineral content and gradually provides growing conditions suitable for the larger grasses and shrubs which eventually choke and shade out the earlier colonisers.

Characteristics of a climax community

A climax community is:

- the **final product** of long-term change within a community;
- **self-perpetuating** and, under natural conditions, not replaced by another community;
- a mature community in **dynamic equilibrium** with its environment.

Effect of environmental factors

Climate

The map in figure 34.2 shows a simplified version of the climatic regions of Western Europe.

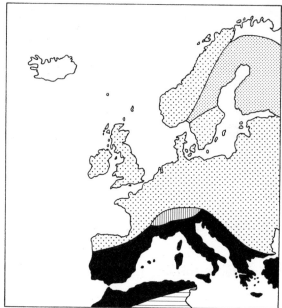

Key to climates		temperature	rainfall
	tundra	very low	medium
	cold temperate	low	high
	cool temperate	medium	high
	mountainous	low	high
	mediterranean	high	low
	desert	very high	very low

Figure 34.2   Climatic regions of West Europe

Despite minor change, climatic conditions such as temperature and rainfall tend to remain fairly constant in a region from year to year. (These may however be permanently disturbed in the future by the greenhouse effect.)

Soil type

The physical, chemical and biological characteristics of a soil are called **edaphic factors**. These include mineral matter, pH and humus content.

The edaphic factors of a soil type found in a particular region are largely determined by the chemical composition of the underlying rock and the prevailing climate affecting the region.

Types of climax community

The types of plant that can survive and the direction taken by succession in a region are dependent upon the climatic and edaphic factors present. The different natural climax

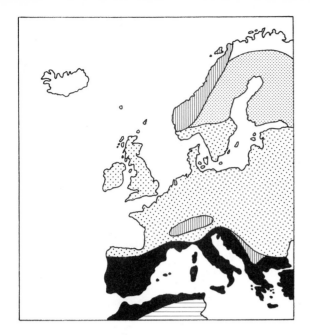

Key to vegetation types

☐ tundra (e.g. lichens)

▦ boreal forest (e.g. conifers)

▦ broad leaf forest (e.g. deciduous trees)

▥ mountain vegetation (e.g. alpine plants)

■ mediterranean scrub (e.g. cacti)

☰ desert vegetation (e.g. thorny bushes)

Figure 34.3 Natural climax communities of West Europe

communities that occur in Western Europe (see figure 34.3) are therefore found to correlate closely with the different climatic regions (figure 34.2).

The **climatic** climax community of a region is the type which is typical of that climatic region and is found growing on the soil type common to the region. In low-lying areas of Britain for example, deciduous forest is the climatic climax community.

An **edaphic** climax community is one which is atypical of the climatic region as a whole and is found growing on a local soil atypical of the climatic region and unable to support the climatic climax community. In Britain, a peat bog of *Sphagnum* moss growing above a waterlogged bed of clay is an example of an edaphic climax community.

Human intervention

Only a few hundred years ago, much of Britain was covered with a climax community of deciduous forest. However most of this has now been cleared to provide land for agriculture, conifer forestry and human settlement. Since this land is rarely left abandoned to nature for extensive periods, succession leading to the climatic climax community of deciduous forest is normally prevented.

Succession leading to edaphic climax communities on less useful land is more common. However even the wetlands (vast expanses of peat, mosses and black pools) of Northern Scotland are now being converted to plantations of conifers and being dug up for their peat.

QUESTIONS

1 a) With reference to plant communities, explain what is meant by the term succession.
   b) Give TWO differences between primary and secondary succession.
2 a) State TWO characteristics of a climax community.
   b) Name the TWO environmental factors which determine the climax community typical of a geographical region.
   c) Give THREE examples of climatic climax communities which occur in Europe.
3 The graph in figure 34.4 represents the process of succession on an abandoned field. The thickness of a line corresponds to a species' relative importance at a given time. A broken line indicates presence of seeds which failed to become established.

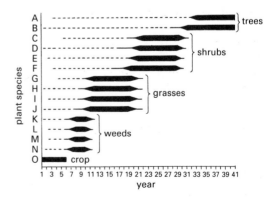

Figure 34.4

199

a) In which year was the field abandoned?
b) (i) How many different species of plant were established in the field in year 20?
(ii) Identify by their letters those grass species which landed on the field as seeds in year 8.
(iii) Of these grass species, how many were represented by seeds that failed in year 8?
(iv) Suggest why these failed.
c) Explain why trees succeeded shrubs.

4 Copy the axes shown in figure 34.5 and draw a line graph to show how biomass varies as succession proceeds.

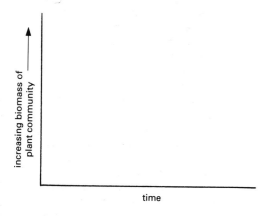

Figure 34.5

DATA INTERPRETATION

Figure 34.6 shows a simplified version of succession on a barren stretch of sand blown up from a sea shore.

a) Does the diagram illustrate an example of primary or secondary succession? Explain your choice of answer. (1)
b) (i) What effect does colonisation by couch grass have on the habitat? (1)
(ii) Which community is eventually able to succeed couch grass? (1)
(iii) What TWO effects does this next community have on the habitat? (1)
(iv) Which community is eventually able to succeed it? Why? (1)
c) (i) Which lettered region on the diagram is able to support the British climatic climax community? (1)
(ii) Some ecologists describe the community of marram grass at region **B** as an edaphic climax community. Is this description justified? Explain your answer. (2)
(iii) Identify another lettered region populated by a community unlikely to be succeeded by the climatic climax if the present conditions persist. (1)

(10)

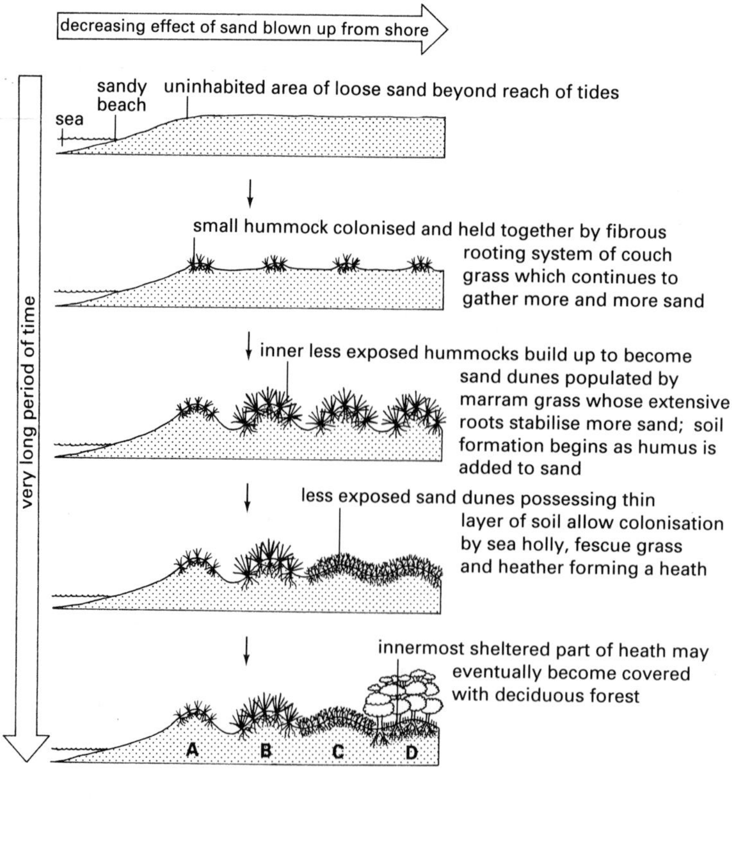

decreasing effect of sand blown up from shore

sandy beach   uninhabited area of loose sand beyond reach of tides

sea

small hummock colonised and held together by fibrous rooting system of couch grass which continues to gather more and more sand

inner less exposed hummocks build up to become sand dunes populated by marram grass whose extensive roots stabilise more sand; soil formation begins as humus is added to sand

less exposed sand dunes possessing thin layer of soil allow colonisation by sea holly, fescue grass and heather forming a heath

innermost sheltered part of heath may eventually become covered with deciduous forest

very long period of time

A   B   C   D

Figure 34.6

What you should know (CHAPTERS 31–34)

1 To function efficiently, many aspects of the human body's **internal environment** must be **maintained** within tolerable limits.

2 **Physiological homeostasis** is the name given to this maintenance of the internal environment despite changes in the external environment.

3 Homeostasis operates on the principle of **negative feedback control**. By this means, a change in the internal environment is detected by **receptors** which send messages to **effectors**. These trigger responses which negate the deviation from the norm and return the internal environment to its **set point**.

4 **Population dynamics** is the study of population changes and the factors which cause them.

5 A population increases until it reaches the **carrying capacity** of the environment and then it remains relatively **stable**.

6 A population is prevented from increasing in size indefinitely by several factors known collectively as **environmental resistance**.

7 Some of these factors affect the growth of the population in a manner **independent** of population density. Other factors operate in a **density-dependent** manner.

8 In a natural ecosystem, a population is kept relatively stable by density-dependent factors effecting **negative feedback control** (homeostasis).

9 The population numbers of many wild plants and animals are **monitored** by humans.

10 This provides information about certain species which: are important as a source of **food**; are **endangered** and in need of protection; act as **indicators** of the environment's health.

11 The change which involves a regular progression from a pioneer community of plants to a climax community is called **succession**.

12 During succession, each community enjoys a period of **short-term stability** during which time it **alters** the habitat. In doing so it makes the habitat less favourable to itself and more favourable to its successor.

13 The **climax community** is the final product of succession whose nature is determined by soil and climatic factors. It enjoys **long-term stability** and is not replaced by another community.

Section 5 Adaptation

35 Maintaining a water balance – animals

Adaptation

An **adaptation** is a characteristic possessed by an organism which makes it well suited to its environment. The members of a species possess certain **evolutionary adaptations** typical of that species. These adaptations are the phenotypic expressions of changes that have occurred in the species' genotype over a very long period of time. They have been favoured by natural selection because possession of them gives the species some advantage and increases its chance of survival.

Some adaptations are **anatomical** (i.e. involve the structure of the organism), some are **physiological** (i.e. depend on the way the organism's body works) and some are **behavioural** (i.e. depend on the way the organism responds to stimuli).

Subsequent chapters will include examples of all three types of adaptation.

Water balance in aquatic animals

Animals such as the jellyfish whose cell contents are **isotonic** (equal in water concentration) to the surrounding environment do not have the problem of maintaining water balance since their cells neither gain nor lose water by osmosis.

A water balance problem does exist however in aquatic animals whose cell contents are **hypertonic** (lower in water concentration) compared with the surrounding environment since they constantly *gain water* by osmosis. Animals whose cell contents are **hypotonic** (higher in water concentration) compared with the surroundings constantly *lose water* to their environment. (These differences in water concentration can also be described in terms of water potential, see Appendix 2).

In each case, these animals possess certain physiological adaptations which enable them to maintain their water (and ion) concentration at the correct level. This control of water (and ion) balance is called **osmoregulation**.

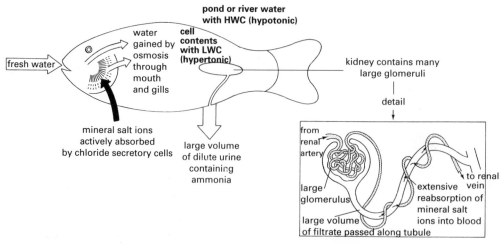

Figure 35.1   Osmoregulation in fresh water bony fish

Osmoregulation in fresh water bony fish

The scaly skin of a fish is impermeable to water. However in a fresh water fish (see figure 35.1), the delicate membranes of the mouth and gills are selectively permeable and water enters continuously by osmosis as the fish breathes. To avoid the body tissues swelling up and becoming damaged, this excess water must be removed.

Over millions of years of evolution, fresh water bony fish have become adapted in several ways to carry out this function. Their kidneys possess **many large glomeruli** (see figure 35.1) which allow a **high filtration rate** of the blood. This results in the production of a very dilute glomerular filtrate. The kidney tubules are very efficient at **reabsorbing mineral salts** (ions) from this glomerular filtrate back into the bloodstream. The fish therefore excretes a **large volume of dilute urine** which contains only a trace of salts and some nitrogenous waste. (Ammonia, the nitrogenous waste, is toxic in high concentrations. However in this type of fish, it is diluted to a harmless level in the urine.)

The membranes of a fresh water bony fish's gills contain **chloride secretory cells** which **actively absorb salts** (ions) from the water as it passes through the gills. The salt concentration of the body is therefore maintained by these cells replacing the small amount lost in the large volume of urine. This process involves **active transport** of ions against a concentration gradient and therefore requires **energy**.

Osmoregulation in salt water bony fish

Although its scaly skin is impermeable to water, this type of fish (see figure 35.2) suffers constant dehydration of its hypotonic tissues. This is because water is continuously lost to the surrounding hypertonic sea water by osmosis through the selectively permeable gill and gut membranes.

This problem is overcome in several ways. **Sea water is drunk** to replace losses. The kidney contains only a **few small glomeruli** (see figure 35.2) or none at all. This results in a **low filtration** rate of blood and only a **little water** being lost in urine. In addition, a salt water bony fish converts its nitrogenous waste to a non-toxic form which requires minimum water for its removal.

The **chloride secretory cells** in the gill membranes work in the reverse direction to those of the fresh water fish. In a marine bony fish they **actively excrete** the excess salt (gained from drinking sea water) back out into the sea. Again this active transport against a concentration gradient requires energy.

Water conservation in a desert mammal

The desert is a region of very low rainfall where drought conditions persist for most of the year (accompanied by high daytime and low night-time temperatures). Hence, desert mammals such as the kangaroo rat have only a very **limited supply of water** available to them. To survive, they have to be able to practise rigorous **water conservation**.

In its natural habitat, the kangaroo rat does not drink water at all. It is able to gain all of its water from its food ('dry' seeds) and remain in water balance as shown in figure 35.3. This is made

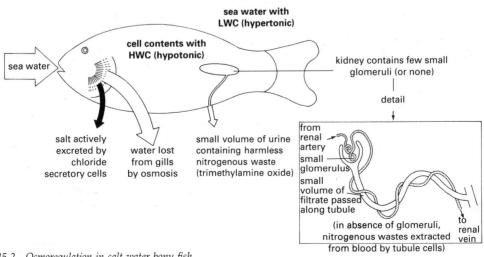

Figure 35.2   Osmoregulation in salt water bony fish

water gain	
6 g in food	
54 g in metabolic water produced during tissue respiration	
60 g = total gain	

water loss	
44 g in expired air from lungs	
13.5 g in urine from kidneys	
2.5 g in faeces from large intestine	
0 g in sweat	
60 g = total loss	

Figure 35.3   *Water balance in kangaroo rat (fed 100 g seeds over one month)*

possible by the possession of certain physiological and behavioural adaptations.

Physiological adaptations

A kangaroo rat's **mouth** and **nasal passages** tend to be **dry**, thereby reducing water loss during expiration.

Its bloodstream contains a **high level of anti diuretic hormone** (see page 174) and its kidney tubules possess very long **loops of Henle**. These adaptations promote water reabsorption so effectively that it can produce urine 17 times as concentrated as its blood. (The most hypertonic urine that humans can produce is only four times as concentrated as blood.)

A kangaroo rat's large intestine is extremely efficient at **reabsorbing water from waste material** and producing faeces with a very low water content.

In addition, a kangaroo rat does not sweat.

Behavioural adaptations

A **behavioural adaptation** is an instinctive behaviour pattern which helps the animal to survive in its natural habitat.

Water conservation is further promoted by the fact that the kangaroo rat remains in its **underground burrow** in a fairly **inactive state** during the extreme heat of the day. Unlike the very hot dry air above ground, the air inside the burrow is **cooler** and **more humid**. Thus the air being inhaled by the rat is almost as damp as the air being exhaled and so minimum net water loss occurs.

The animal is **active at night** and goes out foraging for food when the conditions are cool. It

therefore has no need to produce sweat to give a cooling effect and water is conserved.

Adaptations of migratory fish

A **migratory fish** (e.g. Atlantic salmon and eel) spends part of its life in fresh water and part in salt water and is able to osmoregulate successfully in both environments.

Salmon

The eggs of salmon are laid amongst the gravel in a river bed. After hatching, the young fish spends the first three years of its life feeding and growing until it becomes a **smolt** (the migratory stage). In response to some stimulus (probably a full moon), the smolts migrate downstream together. This is called the **smolt run**.

After lingering for a short time in the river estuary to become **acclimatised** to salt water, the smolts leave the river together and head in vast numbers for their feeding grounds off the west coast of Greenland (see figure 35.4). In these food-rich waters the young salmon increase their body weight by up to 50 times in one year by feeding on shoals of capelin (small herring-like fish).

After one to six years, the salmon respond to some (unknown) stimulus and migrate back across the Atlantic. The returning fish gradually stop feeding, develop reproductive ability and, using sensitive olfactory (smell) receptors, locate the river in which they were originally spawned. The fish are now ready to breed.

On entering the river, they must become acclimatised for a second time to a dramatic

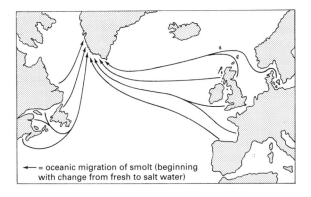

← = oceanic migration of smolt (beginning with change from fresh to salt water)

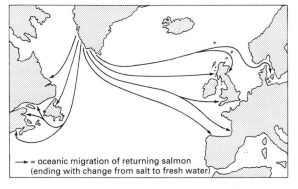

→ = oceanic migration of returning salmon (ending with change from salt to fresh water)

Figure 35.4   Oceanic migrations of Atlantic salmon

change in the water concentration of the external environment.

The precise means by which migratory fish alter their method of osmoregulation is poorly understood but depends at least partly upon the action of several hormones. When a migratory fish moves from salt water to fresh water (and vice versa), its hormonal balance becomes altered. This is accompanied by an appropriate change in glomerular filtration rate and a reversal of the direction in which the chloride secretory cells actively transport salt ions.

This ability to **adapt to changes** in the external environment's water concentration enables migratory fish to exploit a wider range of habitat than fish which are unable to do so.

Modification of a response by an individual

Each individual normally inherits the basic **evolutionary adaptations** (see page 202) typical of the species to which it belongs. Some of these physiological adaptations allow the individual a certain amount of scope within which it can *modify its response to suit the needs of the particular circumstances* that it finds itself in.

For example, all Atlantic salmon possess structurally similar kidneys and chloride secretory cells. However the physiological mechanisms by which these operate from moment to moment and from fish to fish are not rigid and unchangeable. An individual can modify its response to suit a change in the water concentration of its external environment.

Similarly all kangaroo rats possess kidneys with long loops of Henle but the concentration of the urine produced by an individual kangaroo rat will depend upon the availability of water in its environment.

QUESTIONS

---

1 a) With reference to the bar graph in figure 35.5, briefly outline the water balance problem encountered by each type of fish in its natural habitat.

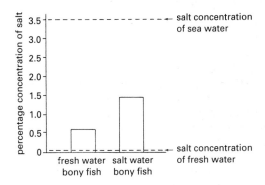

Figure 35.5

b) Copy and complete table 35.1 which refers to adaptations possessed by each of these types of fish to solve its osmoregulatory problems.

	fresh water bony fish	salt water bony fish
relative size of glomeruli		
relative number of glomeruli		
relative filtration rate of blood		
relative volume of urine		
relative concentration of urine		
direction in which salts are actively transported by chloride secretory cells		

Table 35.1

c) State TWO differences between diffusion and active transport.

2 a) Name TWO ways in which a desert rat (i) gains, (ii) loses water.
b) Name TWO physiological and TWO behavioural adaptations shown by a desert mammal which help it to conserve water.

3 Figure 35.6 shows the migratory patterns of eels. Spawning takes place in the sea and, after about three years, the young eels (elvers) travel to the rivers of Europe where they feed and grow to adulthood.

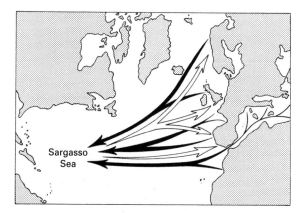

Figure 35.6

When sexually mature, the adults migrate downstream to the sea. They then set off on a journey of thousands of miles to a region of the Atlantic Ocean called the Sargasso Sea where they spawn and begin the life cycle again.
a) Which type or arrow (shaded or unshaded) represents:
(i) the migration of adult eels to their breeding grounds;
(ii) the movement of young eels to the regions where they do most of their feeding and growing?
b) Both of these movements involve a change in environment from one osmotic extreme to another. In each case, describe how the eel adjusts its osmoregulatory system with reference to:
(i) filtration rate of the kidney;
(ii) volume of urine excreted;
(iii) direction of activity of chloride secretory cells.
c) In what way is eel migration different from that of salmon?
d) What benefit is gained by fish able to adapt to changes in the external environment and migrate compared with those unable to do so?

4 Several specimens of two species of crab (**A** and **B**) were submerged in sea water (3.5 per cent salt solution) and given time to acclimatise. A sample of each crab's blood was then taken and analysed to measure its salt content.

The experiment was repeated using 3 per cent salt solution and successively more dilute salt solutions. Average values of blood salt content for both species of crab in each external salt solution were calculated. The graph of the data is shown in figure 35.7.

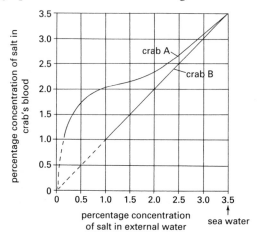

Figure 35.7

a) Which species of crab has blood which is always isotonic to the surrounding water?
b) (i) Which species of crab has blood which remains hypertonic to the surrounding water as the salt content of this external medium decreases below that of sea water?
(ii) At which per cent concentration of salt in the external water is this crab's blood most hypertonic to its surroundings?
c) (i) Which crab is able to exert some degree of osmoregulation?
(ii) Why is oxygen required for this process?
d) Copy and complete table 35.2 by identifying which species of crab is suited to life in each of the habitats and briefly explaining why.

habitat	species of crab	reason
river estuary		
deep sea water		

Table 35.2

36 Maintaining a water balance – plants

Transpiration stream

The continuous passage of water (and nutrient ions) flowing up through a plant from the roots to the leaves is called the **transpiration stream**.

Entry of water

Osmosis

A **root hair** (see figure 36.1) is an extension of an epidermal cell which presents a *large absorbing surface* to the surrounding soil solution. Since the cell sap inside a root hair is a region of lower water concentration than the soil solution outside, water passes into a root hair by **osmosis** through the selectively permeable plasma membrane.

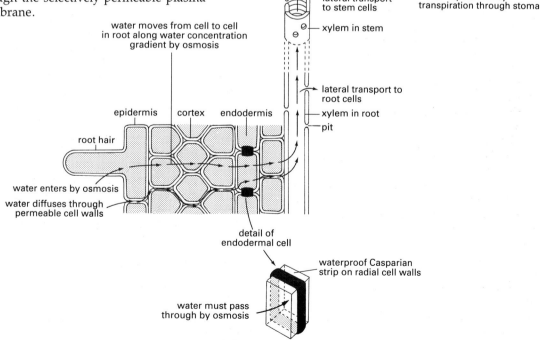

Figure 36.1 Water uptake in transpiration stream

Water then passes from the root hair (which now has a higher water concentration) into a neighbouring cortex cell (which by comparison has a lower water concentration) and so on across the root to the xylem vessels along a **water concentration gradient**. (This movement of water can also be described in terms of water potential, see Appendix 2.)

Diffusion

Many water-filled spaces occur between the cellulose fibres which make up cell walls. Much water therefore **diffuses** across the root via the permeable cell walls and intercellular spaces without entering the living cells.

However this flow of water is interrupted by a layer of cells called the **endodermis**. Their radial walls possess a **waterproof layer**, the **Casparian strip** (see figure 36.1). At this point, the water bypasses the strip by taking the osmotic route through the living cells.

Root pressure

If a plant's stem is severed, the cut stump continues to exude water. The force with which this water is pushed up the stem by the roots is called **root pressure**. It can be measured using a **manometer** as shown in figure 36.2. This movement of water is due to the flow of water by osmosis across the root. It is thought to play only a minor role in the ascent of water via the transpiration stream.

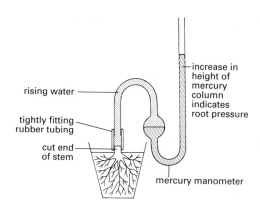

Figure 36.2   *Measuring root pressure*

Capillarity

Adhesion is the force of attraction between *unlike* particles. In the experiment shown in figure 36.3, **capillarity** (force of capillary attraction) occurs because liquid water molecules adhere to the solid glass particles. Water rises to the highest level in the narrowest tube because it presents

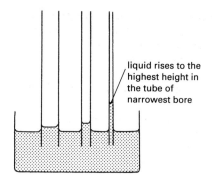

Figure 36.3   *Capillarity*

the largest relative surface area of glass in contact with adhering water molecules.

Since xylem vessels are also tubes of very fine bore, water similarly adheres to the vessel walls. Capillarity is therefore thought to play a minor role in the upward movement of water in a plant.

Upward transport in xylem

Water passes up through the plant in the **xylem** tissue which consists of a continuous system of hollow vessels and tracheids (see page 3) whose structure exactly suits their functions of water transport and support. **Pits** in the walls of these cells allow **lateral transport** of water to other tissues all the way up the plant.

Exit of water

Transpiration is the process by which water is lost by evaporation from the aerial parts of a plant. Most transpiration occurs through holes in the leaves called **stomata**.

The contents of leaf cells which are closest to a moist air space from which water is being lost (via a stoma) have a lower water concentration than cells nearer the xylem vessels. Thus a **water concentration gradient** also exists in a leaf and water passes from xylem to leaf cells by osmosis as shown in figure 36.1. Some of this water is used to maintain cell turgor and as a raw material in photosynthesis.

The walls of the cells in contact with a vein also draw water from the xylem vessels. In fact most of the water that crosses a leaf does so via the cell walls (as in the root) without entering the cells.

Transpiration pull

Water from the walls of the leaf cells lining the moist air spaces evaporates and is lost by transpiration via the stomata. In order to replace these losses, cells draw water from the xylem vessels and set up a **transpiration pull**. This is the *major force* which brings about the ascent of water in the transpiration stream.

Cohesion is the force of attraction between **like** particles. Transpiration pull is explained in terms of the **cohesion-tension theory** which proposes that such raising of water against gravity is possible because water molecules cohere strongly together. This allows thin tense columns of water to be pulled up the xylem vessels to the top of the tallest tree without the water columns breaking.

The diameter of a tree is found to decrease during periods of maximum transpiration. This is thought to occur because the columns of water molecules (drawn thin by the tension exerted by the transpiration pull) still continue to adhere to the xylem vessel walls and pull them inwards.

Other nutrient ions enter by diffusion through the water-filled spaces in the cell walls. However this route is again blocked by the Casparian strips in the endodermis. At this point the ions must be actively transported via the living contents of the cells.

In both cases those mineral ions not retained for use by root cells are **actively pumped** into the xylem vessels and then depend on the transpiration stream for **passive transport** up the plant. Some active transport of ions occurs *laterally* into cambium, phloem and metabolising tissues along the way.

On arriving in the leaf via the transpiration stream, the remaining ions are actively transported into leaf cells. **Iron** and **magnesium**,

Mineral (nutrient) ion uptake

Some nutrient ions enter the plant by **active uptake** (see figure 36.4) and then move from cell to cell via **plasmodesmata** (tiny intercellular cytoplasmic connections, see figure A.1.2 on page 246).

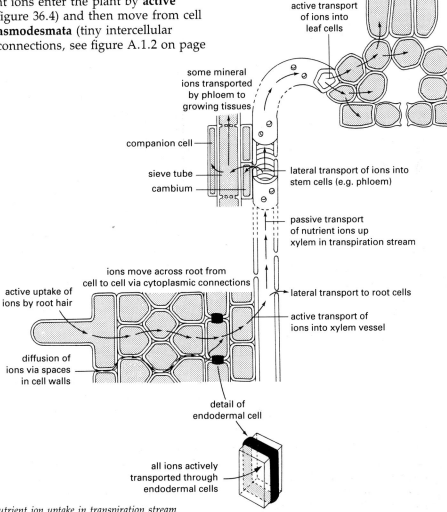

active transport of ions into leaf cells

some mineral ions transported by phloem to growing tissues

companion cell

sieve tube

cambium

lateral transport of ions into stem cells (e.g. phloem)

passive transport of nutrient ions up xylem in transpiration stream

ions move across root from cell to cell via cytoplasmic connections

active uptake of ions by root hair

diffusion of ions via spaces in cell walls

lateral transport to root cells

active transport of ions into xylem vessel

detail of endodermal cell

all ions actively transported through endodermal cells

Figure 36.4 Nutrient ion uptake in transpiration stream

for example, are needed for the formation of chlorophyll.

Importance of transpiration stream

Water and **nutrient ions** are essential to all living things for a variety of purposes. Water is the **universal solvent** in which biochemical reactions occur. In addition green plants need water for **photosynthesis** and the **maintenance of turgor** (especially important for support in non-woody plants).

Certain mineral ions are required by the plant for healthy growth and various metabolic processes (see page 162).

The **transpiration stream** is the source from which cells in all parts of the plant can directly or indirectly gain a supply of water and nutrient ions for these essential functions.

Stomata

Distribution of stomata

In **monocotyledonous** plants (e.g. grass) stomata are found fairly equally distributed on both leaf surfaces. In **dicotyledonous** plants (e.g. geranium) they are almost entirely restricted to the lower epidermis.

Stomatal mechanism

Guard cells differ from normal epidermal cells in three ways. Guard cells are **sausage-shaped** and possess **chloroplasts**. In addition the inner regions of their cell walls (i.e. those facing the stomatal pore) are **thicker** and **less elastic** than the outer regions of their cell walls (see figure 36.5).

Opening and closing of stomata occur as a result of *changes* in *turgor* of guard cells.

When water enters a pair of flaccid guard cells, turgor increases. Due to their larger surface area and greater elasticity, the thin outer parts of the two guard cells' walls become stretched more than the thick inner parts. As the two guard cells bulge out, the thick inner walls become pulled apart *opening* the *stoma*.

When water leaves a pair of guard cells, they lose turgor and return to a flaccid condition. This results in the *closing* of the *stoma*.

Disadvantage of transpiration
Guard cells gain turgor in *daylight* and the stomata open. This physiological adaptation is ideal since the plant needs to take in carbon dioxide for photosynthesis.

However open stomata also allow the loss of

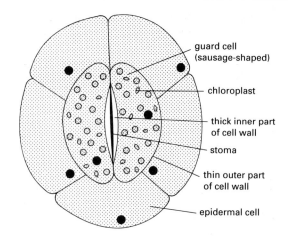

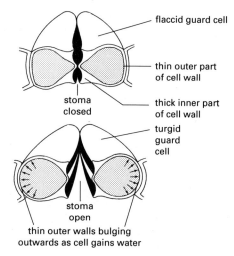

Figure 36.5 Stomata and stomatal mechanism

water by evaporation. Although such transpiration keeps the plant cool in hot weather, excessive transpiration during a dry sunny spell creates a *potential water balance problem*: the amount of water lost at the leaf surfaces may exceed that being absorbed by the roots.

Under such circumstances, leaf cells including guard cells lose turgor. Stomata are therefore often found to close for a time during the day in very hot weather. Since this cuts down water loss, the cells gradually regain turgor and the stomata may reopen later in the day.

Guard cells also lose turgor in *darkness*. Stomata therefore close at night and water is conserved. By controlling its water loss according to external conditions, a plant is capable of a limited degree of **osmoregulation**.

During a very dry spell when stomata are shut, cells still continue to lose turgor because a little water is lost directly through the cuticle of the leaves. At first wilting occurs but, if the drought continues, the plant dies since it is unable to maintain its water balance.

Comparing transpiration rates using a potometer

A **bubble potometer** (see figure 36.6) is an instrument used to measure **rate of water uptake** by a leafy shoot. This rate of water uptake is only approximately equal to transpiration rate since some water may be retained by the leafy shoot for other processes (e.g. photosynthesis).

To set up a bubble potometer, the stem of a leafy shoot is *cut under water* and attached to a length of tightly fitting rubber tubing. With the cut end of the stem and the rubber tubing still immersed, one end of the capillary tubing is inserted into the rubber tubing and gently pushed until water flows along the length of tubing.

The open end of the capillary tube is transferred to the beaker of water and the end lifted to allow an **air bubble** to enter. Once the bubble has appeared on the horizontal arm of the potometer, its rate of movement along the scale is measured (e.g. in mm/min).

The syringe is used to inject water and return the bubble to the start of the scale allowing the experiment to be repeated for this 'normal' environmental condition.

The plant is then subjected in turn to each environmental condition and allowed time to **equilibrate** before its rate of water uptake is measured. Repeat runs are done in each case and the average rate of movement of the bubble is calculated.

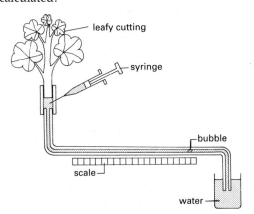

Figure 36.6 Bubble potometer

design feature or precaution	reason
stem cut under water and end of stem, rubber tubing and capillary tubing all connected under water	to prevent air entering xylem and forming air locks
tightly fitting rubber tubing used	to prevent leakage of water and ensure that system is air-tight
time allowed for plant to equilibrate between different environmental conditions	to ensure that rate of movement of bubble is governed by factor being investigated and not the previous one
repeat measurements of rate of movement of bubble taken for each condition and average calculated	to obtain a more reliable result for each condition
all factors kept equal except for one change in environmental conditions	to ensure that only one variable factor is altered at a time

Table 36.1 Experimental design features and precautions

environmental condition	additional apparatus needed to create condition	average rate of movement of bubble (mm/min)
normal day	none	5
windy day	electric fan	20
humid day	transparent plastic bag	1

Table 36.2 Bubble potometer results

Table 36.1 summarises the reasons for adopting certain techniques and precautions during the investigation. Table 36.2 gives a specimen set of results for three different environmental conditions.

Discussion of results

From this experiment it is concluded that **wind** increases rate of transpiration. This occurs because the air outside the stomata is continuously being replaced with drier air which accepts more water vapour from the plant.

Increased **humidity** of the air surrounding the plant results in decreased transpiration rate. This occurs because the concentration gradient of water vapour between the inside and the outside of the leaf is decreased. Rate of diffusion of water molecules therefore slows down.

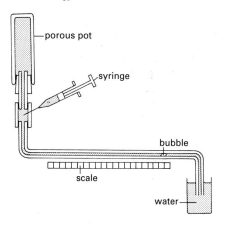

Figure 36.7 *Bubble atmometer*

Bubble atmometer

A **bubble atmometer** (see figure 36.7) is an instrument used to measure **rate of evaporation** from a non-living surface.

When a bubble potometer and a bubble atmometer are subjected to the same conditions, the changes in rate of water loss from a living and a non-living system can be compared. In light for example, both lose water rapidly but in darkness only the atmometer continues to do so since the plant has closed its stomata.

Thus the presence of an atmometer shows when the potometer is acting as a free evaporator and when it is affected by **physiological** factors such as stomatal closure.

Further factors affecting transpiration rate

In addition to light, availability of soil water, humidity of the surrounding air and wind speed, transpiration rate is also affected by the following factors.

Temperature
Transpiration increases with increase in temperature due to the faster evaporation rate of water molecules.

Air pollution
Transpiration rate is decreased if the stomata are blocked with dirt.

Xerophytes

'Normal' plants, which live in habitats where water is abundant and excessive transpiration does not occur, are called **mesophytes**.

Xerophytes are plants which live in habitats where a mesophyte would not survive because its transpiration rate would be excessively high. Such habitats are characterised by either hot, dry conditions and lack of soil water (e.g. desert) or exposed, windy conditions (e.g. moorland).

Xerophytes are able to *maintain a water balance* in such extreme habitats because they have evolved certain **xeromorphic adaptations** as follows.

Structural adaptations which reduce transpiration rate

Water loss is cut to a minimum in leaves which possess a **reduced number of stomata** and are covered by a **thick cuticle**. Transpiration is further reduced if the leaf is **rolled** and/or **hairy** (see figure 36.8) since each of these adaptations traps a layer of moist, relatively immobile air between the stomata and the outer atmosphere. **Stomata sunken in pits** (see figure 36.9) are similarly protected because the pits trap pockets of moist air.

Some leaves (e.g. pine needles) are **small** and **circular** in cross-section thereby reducing the relative surface area of transpiring leaf exposed to the atmosphere.

In many cacti, the leaves are reduced to

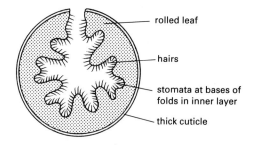

Figure 36.8 *Transverse section of marram grass leaf*

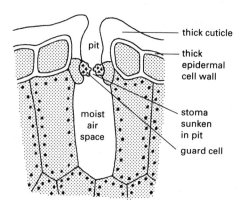

Figure 36.9 *Detail of* Hakea *leaf*

protective **spines** and water loss is limited to the stem which carries out photosynthesis and possesses relatively few stomata.

Structural adaptations for resisting drought

Many cacti have **long roots** allowing absorption of subterranean water. Others possess extensive systems of **superficial roots** which grow parallel to the soil surface enabling them to absorb maximum water on those rare occasions when rain does fall.

Cacti store this water in **succulent tissues** (see figure 36.10) and may even have a **folded stem** which allows expansion and contraction (subject to water availability) without becoming damaged.

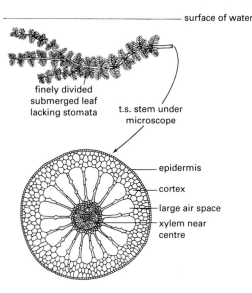

finely divided
submerged leaf
lacking stomata

t.s. stem under microscope

surface of water

epidermis

cortex

large air space

xylem near centre

Figure 36.11   Water-milfoil

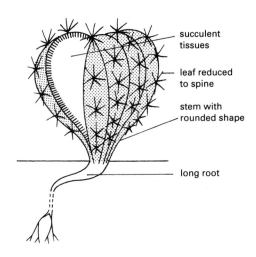

succulent tissues

leaf reduced to spine

stem with rounded shape

long root

Figure 36.10   Cactus plant (with part of stem cut out)

Physiological adaptations

Most cacti reduce water loss by showing **reversed stomatal rhythm** (i.e. closed during the day and open at night). During the night carbon dioxide is taken in and stored for use in photosynthesis in the daytime when the stomata are closed.

Some xerophytes evade drought by **ceasing vegetatative activity** during dry times. For example they survive in a highly **desiccated** state inside a hard seed coat and only germinate and grow when water becomes available.

Hydrophytes

Hydrophytes are plants which live completely submerged (e.g. water-milfoil) or partially submerged (e.g. water-lily) in water. They have evolved many adaptations which help them to survive in their aquatic environment.

Air spaces

Although gaseous exchange occurs all over its surface, a submerged plant is faced with the problem of obtaining an adequate supply of oxygen for respiration since this gas is only sparingly soluble in water.

A hydrophyte is adapted to overcome this problem by possessing an extensive system of intercommunicating **air-filled cavities** (see figure 36.11) throughout its submerged organs. Instead of escaping into the surrounding water, much of the oxygen formed during photosynthesis is stored in these air spaces ready for use in respiration when required.

The presence of such **aeration tissue** also gives a submerged plant buoyancy in order to keep its leaves near the surface for light.

Reduction of xylem

Since water provides a submerged plant with support and is readily available for absorption all over its surface, a hydrophyte is found to possess little strengthening or water-conducting tissue. Any xylem present is normally found at the *centre* of the stem (see figure 36.11). This allows the stem **maximum flexibility** in response to water movements while at the same time enabling it to **resist pulling strains**.

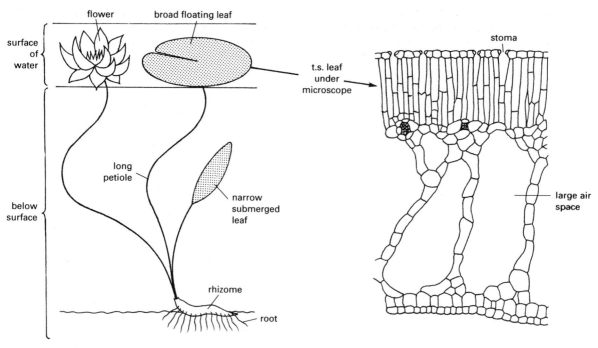

Figure 36.12   Water-lily

Specialised leaves

A hydrophyte's submerged leaves are **narrow** in shape or **finely divided** (see figure 36.11). This adaptation helps to prevent them from being torn by water currents.

Stomata must be in contact with the air to bring about gaseous exchange. Submerged leaves therefore lack stomata and floating leaves have all their stomata on their *upper* surfaces (see figure 36.12). Such floating leaves often possess **long leaf stalks** (petioles) which prevent the stomata from being flooded when the water level rises.

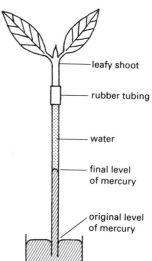

Figure 36.13

QUESTIONS

1 a) Starting in the soil solution, describe a possible route through a plant that could be taken by:
(i) a molecule of water which is eventually lost by transpiration;
(ii) An ion of magnesium which is eventually used in the formation of chlorophyll.
b) What name is given to the passage of water and nutrient ions up through a plant?
2 a) Explain why the stem of the leafy shoot shown in the experiment in figure 36.13 must be cut under water.
b) State a further precaution which must be adopted when setting up this experiment.
c) To what important theory does the result obtained in this experiment lend strong support? Explain your answer.
3 a) Describe what happens to the tree's diameter during the 24 hour period shown in the graph in figure 36.14.
b) Briefly explain why the process of transpiration is held responsible for this effect.

214

4 a) Describe the mechanism by which a plant is able to osmoregulate in response to temporary shortage of soil water during a hot dry day.
b) Under what circumstances could such maintenance of a water balance break down? Explain why.

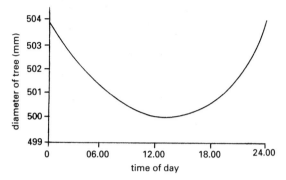

Figure 36.14

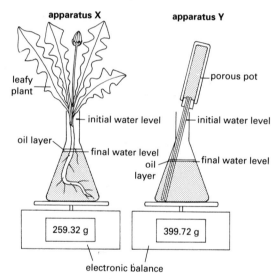

Figure 36.15

	apparatus X	apparatus Y
initial mass (g)	283.80	
mass after 2 days (g)		
mass of water lost (g)		56.00
volume of water lost (cm³)	24.48	56.00
volume of water required to restore initial level (cm³)	25.00	56.00

Table 36.3

5 The readings on the balances in figure 36.15 show the final masses of two sets of apparatus (**X** and **Y**) which have been allowed to lose water for two days in a laboratory.
The partly completed table 36.3 records the initial mass of apparatus **X** and the volume of water which had to be added to restore the water in each flask to its initial level.
a) Copy and complete this table.
b) Calculate the average mass of water lost per hour by the plant.
c) Explain why the volume of water needed to restore the initial level in **X** differed from the volume lost by the plant.
d) Predict the effect on water loss of subjecting **X** and **Y** to the following conditions for two days:
(i) covering each with a transparent plastic bag;
(ii) keeping each in the same laboratory in total darkness.
Explain your answer in each case.
e) (i) Apparatus **X** is called a weight potometer. Name apparatus **Y**.
(ii) Why is apparatus **Y** included in this experiment?

6 Four large leaves of similar size were removed from a horse chestnut tree and treated as follows:

leaf **A**   vaseline applied to both surfaces
leaf **B**   no vaseline applied
leaf **C**   vaseline applied to lower surface only
leaf **D**   vaseline applied to upper surface only

Each leaf was then attached to a bubble potometer of uniform design. The four potometers were kept in identical environmental conditions and the distance travelled by the bubble recorded for each over a period of ten minutes. The results are shown in table 36.4.
a) Present the data as line graphs.
b) Which leaf showed the greatest rate of transpiration?
Explain why.
c) Which leaf would be unable to lose water by transpiration?

time (in min)	total distance travelled by bubble (mm)			
	leaf **A**	leaf **B**	leaf **C**	leaf **D**
start	0	0	0	0
2	1	30	6	20
4	2	80	12	50
6	3	128	18	92
8	3	168	24	130
10	3	200	30	158

Table 36.4

d) What information in the table justifies the claim that the rate of movement of the bubble is not exactly equal to the rate of transpiration?
e) What does the movement of the bubble in a potometer actually measure?
f) Which surface of this type of leaf possesses more stomata?
Explain how you arrived at your answer.
7 a) Name THREE xeromorphic adaptations shown by Scots pine as illustrated in figure 36.16.

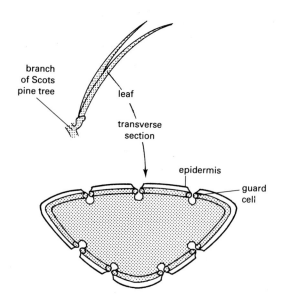

Figure 36.16

b) Explain briefly how each of these features is related to the reduction in rate of water loss.
8 a) Give TWO reasons why hydrophytes normally possess very little xylem tissue compared with land plants.
b) State TWO benefits gained by hydrophytes which possess large air spaces amongst their tissues.

EXPERIMENTAL DESIGN QUESTION

Observations: Cobalt chloride paper is blue when dry and pink when dampened by water vapour. Blue cobalt chloride paper attached to the upper and lower leaf surfaces of an iris plant was found to turn pink in both cases within a similar length of time.

Hypothesis: The number of stomata present on both the upper and lower surfaces of an iris leaf is approximately equal.

Problem: Design an experiment which would enable you to determine the rate at which water is lost by transpiration from each leaf surface, thereby testing the validity of the hypothesis above.

DATA INTERPRETATION

The data in table 36.5 refer to a series of experiments in which a large leafy shoot attached to a bubble potometer was subjected to various conditions of several abiotic factors.

Table 36.6 records the results from a further series of experiments in which the effect of a wider range of one of the abiotic factors was studied.
a) Consider each of the following pairs of experiments in turn; for each pair draw a conclusion about the effect of a named abiotic factor on the time taken by the bubble to travel 100 mm:
(i) **1** and **2** (ii) **3** and **7** (iii) **6** and **10**. (3)
b) Why is it impossible to draw a valid conclusion about the effect of an abiotic

		experiment number									
		1	2	3	4	5	6	7	8	9	10
abiotic factor	wind speed (m/s)	0	0	0	0	15	15	15	15	15	15
	temperature (°C)	5	25	5	25	5	25	5	25	5	25
	air humidity (%)	75	75	95	95	75	75	95	95	75	75
	light (L)/dark (D)	L	L	L	L	L	L	L	L	D	D
time taken by bubble to travel 100 mm		3 min 3 s	1 min 35 s	4 min 28 s	3 min 2 s	1 min 56 s	0 min 30 s	3 min 22 s	1 min 57 s	24 min 10 s	22 min 4 s

Table 36.5

		experiment number				
		11	**12**	**13**	**14**	**15**
abiotic factor	wind speed (m/s)	5	5	5	5	5
	temperature (°C)	20	20	20	20	20
	air humidity (%)	75	80	85	90	95
	light (L)/dark (D)	L	L	L	L	L
time taken by bubble to travel 100 mm		1 min 34 s	1 min 56 s	2 min 20 s	2 min 40 s	3 min 2 s

Table 36.6

factor on time taken by the bubble from a comparison of experiments **3** and **8**? (1)

c) Which experiment in table 36.5 should be compared with experiment **5** to ascertain the effect of darkness on time taken by the bubble to travel 100 mm? (1)

d) Which TWO experiments should be compared in order to find out the effect of wind speed on time taken by the bubble when the plant is in light at 25°C and in air of 95% humidity? (1)

e) Which TWO experiments should be compared in order to find out the effect of

temperature on transpiration rate by the plant when exposed to windspeed of 15 m/s in light and in air of 75% humidity? (1)

f) (i) Name the one variable factor studied in the series of experiments given in table 36.6.

(ii) State the relationship that exists between this variable factor and transpiration rate.

(iii) Account for this relationship in terms of water vapour molecules. (3)

(10)

What you should know (CHAPTERS 35–36)

1 **Fresh water** bony fish constantly **gain water** by osmosis from the surrounding hypotonic water.

2 Their kidneys possess **many large glomeruli** which allow a high filtration rate of blood resulting in the loss of much water in urine.

3 Their **chloride secretory cells** actively **absorb salts** from external water.

4 **Salt water** bony fish constantly **lose water** by osmosis to the surrounding hypertonic water.

5 They drink sea water and have kidneys with **few small glomeruli** which allow only a little water to be lost in urine.

6 Their chloride secretory cells actively **excrete excess salt** out into the sea.

7 **Desert** mammals show **physiological** and **behavioural adaptations** which enable them to conserve water.

8 **Migratory** fish are adapted to cope with changes in water concentration of the external environment by being able to **alter** their **method of osmoregulation** as required.

9 The continuous passage of water and nutrient ions up through a plant is called the **transpiration stream**. It is the means by which cells are supplied with ions and water needed for synthesis reactions.

10 The force with which water is pushed up a stem by the roots is called **root pressure**.

11 **Transpiration** is the loss of water by evaporation. It occurs mainly through stomata in leaves.

12 This loss of water sets up a **transpiration pull** which draws columns of water molecules up through the xylem vessels. This is explained in terms of the **cohesion-tension theory**.

13 **Changes in turgor** of guard cells bring about **opening** and **closing** of stomata.

14 Transpiration **rate** is affected by several different environmental factors.

15 **Xerophytes** are plants adapted to survive in dry or exposed habitats.

16 **Hydrophytes** are plants adapted to survive completely or partially submerged in water.

37 Obtaining food – animals

When animals go searching (**foraging**) for food, they show distinct behaviour patterns organised to gain maximum energy.

Investigating planarian activity

Planaria (see figure 37.1) is a type of flatworm. Since it is muscular it is able to move freely about and change direction.

In the experiment shown in figure 37.2, ten hungry planarian worms of the same species are placed in both petri dishes which contain clean aerated pond water at room temperature. The animals are left to move about at random and become acclimatised for a few minutes.

A tiny piece of fresh raw liver is then placed at the centre of dish **A** and a glass bead (similar in size to the liver) is placed at the centre of dish **B** (the control).

After a few minutes most or all of the *Planaria* in dish **A** are found to have moved directly towards, and begun feeding on, the liver. In dish **B** the animals remain distributed at random. Repeats of the experiment are found to give the same results.

It is concluded from this experiment that planarian worms are able to sense and move directly towards a suitable foodstuff.

Since this experiment is carried out in light it could be argued that the animals use their eyes to spot the food. However when the experiment is repeated in darkness, the animals are equally successful. This shows that they employ **chemoreceptors** to forage for food.

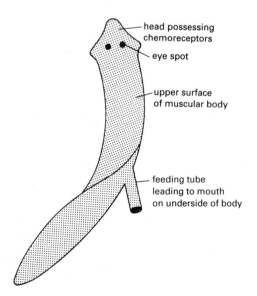

Figure 37.1   Planaria

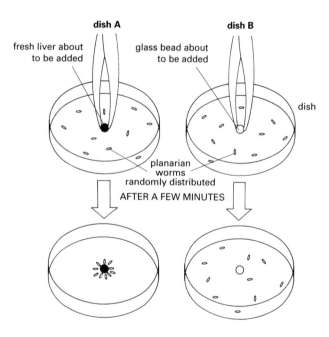

Figure 37.2   Planarian activity in presence of food

Foraging behaviour and search patterns

Bees

Bees are social insects. The survival of a colony is dependent on a regular food supply. Bees go foraging amongst flowers for food. Once a worker has located a good food source, she returns to the hive and performs a **waggle dance** which indicates the *distance* and *direction* of the food source to other workers.

Since the duration of the dance is directly related to the amount of food present at the indicated source, the attention of newly recruited foragers is kept focused on those locations offering most food.

Ants

When ants go foraging, each covers an area of ground and often 'meanders' back and forth over it. This search pattern increases the animal's chance of finding food as close to the colony as possible.

Once food has been located, the successful forager leaves **scent marks** on the ground as it heads directly back to the colony. Other ants quickly follow the trail to the food and reinforce the scent marks as long as the food supply lasts. Since the scent marks quickly fade, no energy is wasted following an old trail that no longer leads to food (see figure 37.3).

Higher animals

The foraging information contained in the bee dance and the ant trail are social communication devices based on instinctive behaviour. Higher animals also forage for food. However their search patterns are normally more complex since their behaviour is not purely instinctive but also involves aspects of **learning**.

Amongst predatory animals (which have to hunt and catch prey) a vast variety of methods of obtaining food exists. These often involve sophisticated physiological adaptations being combined with behavioural adaptations.

A bat, for example, emits **high-pitched sounds** which bounce off objects in the immediate environment. These echoes are picked up by the bat's ears enabling it to locate prey in the dark while at the same time avoiding harmful obstacles.

Whereas a cheetah depends on its **speed** (95 km/h over short distances) to catch its prey, a tiger uses **stealth** during the first part of the hunt and then captures its prey by leaping at it from a distance of up to six metres.

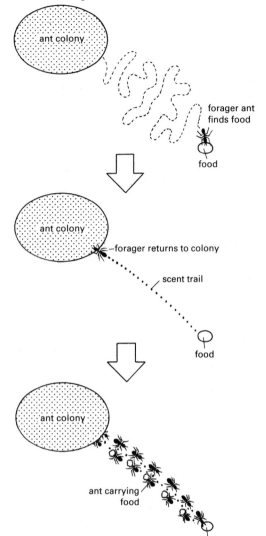

Figure 37.3   Foraging behaviour in ants

Economics of foraging behaviour

Time and energy are consumed by an animal during its search for food. If the energy gained from the food that it finds is less than the energy expended on the search, then the animal suffers a **net loss** of energy.

Such uneconomical behaviour would soon result in death if it were repeated on a regular basis. To be economical an animal must forage **optimally**. This means that it must consume those food items which will give it the best return for the time and energy spent. Foraging behaviour and subsequent choice of food items are therefore affected by the following factors.

Time

The **search time** is the name given to the amount of time spent *locating* the food. The **pursuit time** is the time spent *obtaining* the food once located.

Where the search time is short and the pursuit time long (e.g. a lion in constant sight of a herd of gazelles which are difficult to catch), then it is economical for the predator to be selective and wait for a chance to pick off an old or weak prey animal.

Where the search time is long and the pursuit time short (e.g. a bird searching at length through dense foliage and eventually finding a rich supply of insects which are easy to capture), then it is economical to be non-selective and eat as many prey items as possible since it may take a long time to find a similar haul.

Unproductive versus productive ecosystem

Consider a poor ecosystem where a number of different foods are thinly dispersed. The forager cannot afford to be too choosy because if it ate only the most desirable food items then too much time and energy would be spent on the search. On the other hand if it ate only the poor items, it would not gain enough energy. To be economical it has to settle for a mixture of items with those of intermediate quality giving the best **net energy return** for the time spent on the search (see figure 37.4).

In a rich ecosystem, search time is reduced and the forager can afford to be more selective since it can obtain all of its energy requirements from a few types of choice food items.

Risk versus good food supply
In some cases it is economical for a herbivore to settle for the food available in a poorer ecosystem

if foraging in a food-rich area exposes it unduly to risk of attack by predators.

Size of prey

At first, **net energy gain** (benefit − cost) increases with increasing size of prey since larger items contain more energy than smaller ones. However an optimal prey size is eventually reached as shown in the graph in figure 37.5.

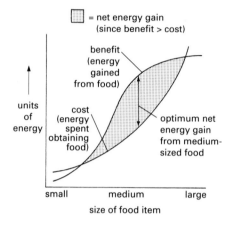

Figure 37.5   Effect of food size on economics of foraging

Beyond this point, net energy gain decreases because the very largest prey is the scarcest (hence involves a longer search time) and it tends to put up the best fight. It is not economical if the predator has to expend much energy subduing the prey.

Competition

Animals compete with one another for resources (such as food, water and shelter) if these are in short supply. Competition between individuals of different species is called **interspecific competition** whereas that between members of the same species is called **intraspecific competition**.

Interspecific competition

When two different species occupy the same **ecological niche** (see page 121), interspecific competition may become so fierce that one species ousts the other. This is called the **competitive exclusion principle**.

Paramecium
Paramecium is a unicellular organism (see page 2). Two closely related species of this animal are

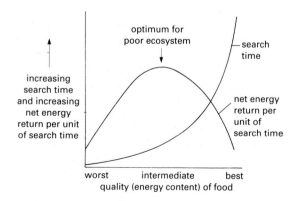

Figure 37.4   Effect of poor ecosystem on economics of foraging

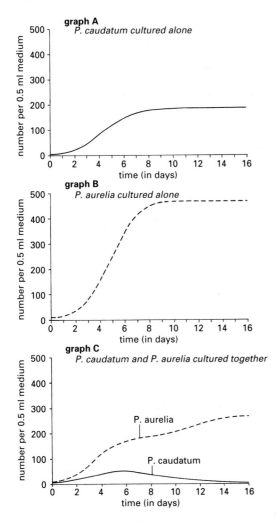

graph A
P. caudatum cultured alone

graph B
P. aurelia cultured alone

graph C
P. caudatum and P. aurelia cultured together

P. aurelia

P. caudatum

Figure 37.6   Effect of interspecific competition on population growth

Paramecium caudatum and *Paramecium aurelia* (which is smaller in cell size).

Graph **A** in figure 37.6 represents the growth of a population of *P. caudatum* cultured in favourable conditions and fed on bacteria. Graph **B** represents the growth of a population of *P. aurelia* cultured separately under exactly the same conditions. Each graph shows a normal population growth curve.

Graph **C** shows the growth curves that result when the two species are cultured together. During the first few days while food is still plentiful, both species increase in number in the absence of competition. However after about six days, the population of *P. caudatum* starts to decline whereas that of *P. aurelia* continues to rise. Eventually, as a result of interspecific competition, *P. caudatum* is completely ousted.

It is possible that *P. aurelia* is the successful competitor because it feeds more efficiently or survives on less food. However, it is interesting to note that *P. aurelia* is also affected by the competition in that its population fails to reach as high a maximum size as it does when cultured alone.

Squirrels
It is thought that the introduction to Britain of the North American grey squirrel which occupies the same ecological niche as the red squirrel, is responsible for the widespread decrease in the numbers of the red squirrel sighted in Britain in recent years.

Cormorants
The common cormorant and the green cormorant are two species of sea-bird which nest on cliffs and dive into the sea for fish. These appear therefore to occupy the same ecological niche. However one species has not ousted the other because competition is minimised by the common cormorant feeding mostly on flatfish and crustaceans from the sea bed whereas the green cormorant feeds further out to sea on fish in the upper water.

It is highly likely, of course, that either type of cormorant, in the absence of the other, would extend its range.

Intraspecific competition

Whereas different species of an animal community can reduce competition by eating different foods, nesting in slightly different habitats, seeking food at different times of the day etc., the members of a population of the same species need *exactly the same resources*.

Intraspecific competition is therefore even more intense than interspecific competition when there is a scarcity of some resource such as food.

Territorial behaviour
Intraspecific competition often takes the form of **territoriality**. This is the name given to behaviour which involves competition between members of the same species (especially birds) for territories.

An animal's total **range** is the area which it covers during its lifetime. Within this range, a male animal often establishes a smaller area called its **territory**. This contains enough food for himself and eventually a mate and their young.

He defends his territory fiercely using **social signals** (sign stimuli) as shown in figures 37.7 and 37.8. He is most aggressive at the centre of his territory. The further he moves away from the centre, the less likely he is to attack intruders.

Eventually during such a contest, there comes a point when he is equally likely to fight or turn and flee.

Such points of balance established by 'pendulum' fighting (alternating attack and

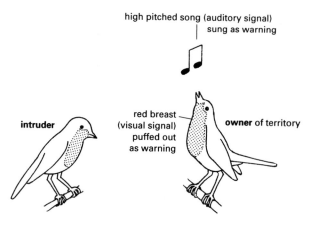

Figure 37.7 Territorial behaviour in robins

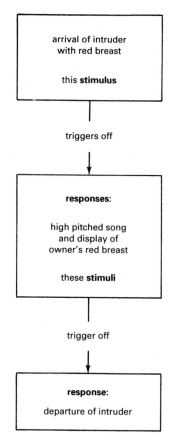

Figure 37.8 Flow chart of territorial behaviour

escape) mark the **boundary** between the territories of two rivals.

Unlike the members of a dominance hierarchy (see opposite) which live in a social group based on rank order, most territory holders are *solitary*. Each enjoys complete dominance within the boundaries of his own territory.

Advantages of territorial behaviour
In addition to providing the animal with a safe place to breed, territorial behaviour *spaces out* the population in relation to the available food supply. This ensures that there will be enough food for the number of young produced. Whereas a bird of prey will defend a full square mile or more, the needs of a robin are met by one small garden.

Red grouse
Within a species, **territorial size** varies depending on the availability of food. The red grouse lives on moorland and feeds on the shoot tips and flowers of heather plants. Young heather plants are more nutritious than older ones. Rather than compete directly for food, the male red grouse claims a territory large enough to provide food for his dependents during the breeding season.

Each enclosed space in figure 37.9 represents a red grouse's territory on the same piece of moorland over a period of four years. During years 1 and 4, food was plentiful and the birds only needed to defend small territories. During years 2 and 3, the heather was poor and a larger territory was required to supply a bird's needs. When times were lean, intraspecific competition was more intense and weaker birds which failed to establish a territory did not breed.

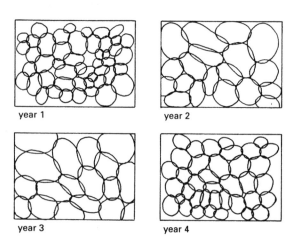

Figure 37.9 Sizes of red grouse territories

Dominance hierarchy

Many higher animals live in large **social groups**. Within such a group, the members react to **social signals** (sign stimuli) given by other members of the same species.

Birds

If a group of newly hatched birds such as pigeons are kept together, one will soon emerge as the **dominant** member of the group. This bird is able to peck and intimidate all other members of the group without being attacked in return. It therefore gets first choice of any available food.

Below this dominant bird there is a second one which can peck all others except the first and so on down the line. Such a linear **peck order** is an example of dominance hierarchy.

Mammals

Although not so clear cut, a similar system of social organisation exists amongst some mammals such as wolves and baboons. Because of his rank, the dominant individual has certain rights such as first choice of food, preferred sleeping places and available mates. In wolves, the dominant male asserts his rank by employing social signals as shown in figure 37.10.

The dominant wolf's **visual display** of ritualised threat gestures is normally enough to assert his authority over other members of the social group. These in turn demonstrate their acceptance of his status by showing **subordinate** responses.

Such a system of hierarchy increases the species' chance of survival since real aggression is kept to a minimum, energy is conserved, experienced leadership is guaranteed and the

Figure 37.11 Co-operative hunting

largest, strongest animals are most likely to pass their genes on to the next generation.

Co-operative hunting

Some predatory mammals such as lions, wolves and wild dogs rely on **co-operation** between members of the social group to hunt their prey.

The stratagem employed by lions, for example, involves some predators driving prey towards others that are hidden in cover and ready to pounce. Dogs and wolves on the other hand take turns to run down a solitary prey animal to the point of exhaustion and then attack it (see figure 37.11).

When a kill is achieved, all members of the predator group bolt down large amounts of meat (some of which is later disgorged by females to feed young). Thus **co-operative hunting** benefits the subordinate animals as well as the dominant leader of the group. By working together, the animals all gain more food than they would by foraging alone. Provided that the food reward gained by co-operative hunting exceeds that from foraging individually, the social group will continue to share food.

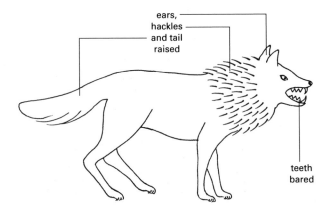
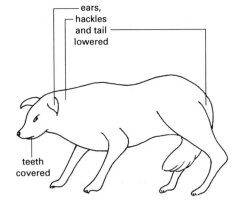

ears, hackles and tail raised

teeth bared

ears, hackles and tail lowered

teeth covered

Figure 37.10 Dominant and subordinate responses

QUESTIONS

1 a) What is meant by the term foraging?
b) *'To be economical, an animal must forage optimally.'*
Briefly explain why.

2 Figure 37.12 shows an experiment set up to investigate the effect of a chemical extracted from the glands of ants on their food-searching behaviour.

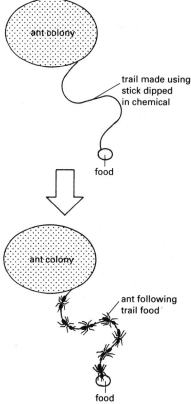

Figure 37.12

a) (i) Describe the control that should have been set up.
(ii) Explain clearly why the control is necessary.
b) What conclusion can be drawn from the experiment (assuming that the control remained unchanged)?
c) What experiment (and control) could be set up to investigate if the presence of food is necessary for ants to follow the chemical trail?
d) The effect of the chemical is short-lived yet more and more ants emerge and follow the trail (even in darkness) provided that the food supply does not run out. How do they find their way?

feature of experimental design	reason
same number of animals of same species used and identical environmental conditions maintained in each dish	
	to ensure that the results are not based on only one animal whose behaviour might not be typical of the species in general
hungry animals used	
	to ensure that the animals are not simply attracted to any physical object
experiment repeated many times	

Table 37.1

e) By what means are wasted journeys to locations that no longer contain food prevented?
3 Copy and complete table 37.1 which refers to some of the design features which apply to the planarian investigation described on page 000.
4 In an investigation involving bats, two environmental chambers, X and Y, were used which contained lengths of wire strung from ceiling to floor and several moths (prey). Hungry blindfold bats were released into chamber X and hungry blindfold bats with their ears plugged were released into Y.
a) State FOUR factors which would have to be kept constant between X and Y to make this a fair test.
b) Identify the variable which was altered.
c) Predict, with reasons, the outcome of the experiment.
5 Instead of fighting, wolves perform a ritual.
a) Describe FOUR features of dominant wolf's visual display when asserting his authority.
b) Describe the corresponding responses displayed by a subordinate animal.
c) What name is given to the type of social organisation that results from this behaviour pattern?
d) Of what advantage is it to the animals concerned?
6 a) Explain the difference between the terms **territory** and **territoriality**.
b) Briefly describe an example of territorial behaviour with reference to a named animal.
c) State TWO advantages of territoriality.

7 Shore crabs feed on mussels by breaking open their shells and eating the soft inners. The graph in figure 37.13 shows the results of offering crabs an unlimited choice of mussels of different sizes.

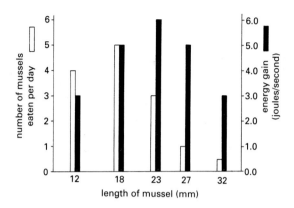

Figure 37.13

a) State the energy gained by a crab consuming:
(i) 5 mussels of length 18 mm per day;
(ii) 3 mussels of length 23 mm per day;
(iii) 1 mussel of length 27 mm per day.
b) Which of these daily diets yields optimum energy?
c) Suggest why tackling the largest size of mussel is uneconomical in terms of energy gain.

DATA INTERPRETATION

Table 37.2 gives the results of a series of competition experiments involving two species of flour beetle, **A** and **B**. The beetles were reared together in containers of flour maintained at the environmental conditions shown in the table. The experiments were run until only one species of beetle ('the winner') remained in each container.

The graphs in figure 37.14 refer to a further experiment using species **A** reared with a much smaller beetle, species **C**, in containers of flour in which a number of short lengths of glass tubing had been buried.

Graph **1** shows the results using glass tubing narrower in diameter than the body size of species

	percentage number of experiments won by species:	
environmental conditions	A	B
dry and cold	100	0
dry and warm	85	15
dry and hot	88	12
wet and cold	69	31
wet and warm	16	84
wet and hot	0	100

Table 37.2

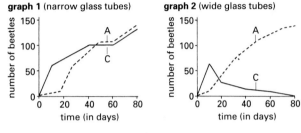

Figure 37.14

A but wider than that of species **C**.

Graph **2** shows the results using glass tubing wider than the body sizes of both **A** and **C**.
a) From the table, state the most favourable environmental conditions for each species. (2)
b) Under what set of conditions was competition probably most intense? Explain your answer. (2)
c) In what way does the data in the table support the principle of competitive exclusion? (1)
d) Predict the outcome of keeping species **A** and **B** in a flour container receiving 12-hourly spells of dry and cold conditions alternated with 12-hourly spells of wet and hot conditions. (2)
e) Compare graphs **1** and **2** with respect to the effect of width of glass tubing on number of species **C**. Give a possible explanation for this difference. (2)
f) Is the competition illustrated by the above experiments intraspecific or interspecific? (1)

(10)

38 Obtaining food – plants

Sessility and mobility

Movement is a characteristic of all living things. An animal can normally move its whole body from place to place and is said therefore to be **mobile**. The movements of higher plants, on the other hand, are restricted to certain parts such as leaves turning towards the light. The plant cannot move about as a complete organism. It remains fixed to one position and is said to be immobile or **sessile**.

Such mobility and sessility have a bearing on the strategies employed by animals and plants to obtain food.

Heterotrophic nutrition

Animals are **heterotrophic**. This means that they need a ready-made organic substance to provide them with energy and materials for building cytoplasm. They are therefore dependent, directly or indirectly, on the synthetic activities of green plants.

In order to obtain food, most animals have to go **foraging** (see chapter 37) and so they must be mobile. This need for mobility puts a limit on the maximum body size that an animal can attain and still function efficiently.

Autotrophic nutrition

Green plants are **autotrophic**. This means that they do not depend on an outside source of organic food but are able to synthesize all of their organic requirements from simple inorganic molecules. During **photosynthesis**, carbon dioxide and water are used to produce carbohydrates. Nitrate absorbed from the immediate environment provides the nitrogen needed for protein synthesis.

Since all the necessary chemicals are 'to hand', green plants produce their own food 'on site'

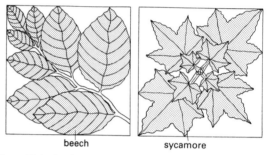

beech   sycamore

Figure 38.1   Leaf mosaic patterns

and have no need to be mobile. This sessile existence allows many plants to attain a body size greatly in excess of any animal.

Leaf mosaic patterns
Natural selection favours a characteristic which increases the amount of light that a plant is able to absorb. A plant's leaves are therefore often arranged in a **mosaic pattern** (see figure 38.1).

This allows the maximum surface area of leaf to receive light since overshadowing of one leaf by another is reduced to a minimum. The leaf mosaic pattern of beech leaves is so effective that almost no light passes through the leaf canopy of a beech wood and only a few shade-tolerant species (see page 229) can survive on the ground below.

Competition between plants

Plants growing in the same habitat *compete* with one another for factors such as light, water and soil nutrients if any one of these is in short supply.

Investigating intraspecific competition

Plants of the same species have exactly the same growth requirements. When grown together, they will be in direct competition with one another if any resource is *limiting*.

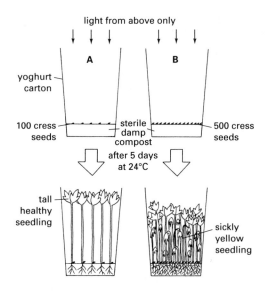

light from above only

A   B

yoghurt carton

100 cress seeds ——— sterile damp compost ——— 500 cress seeds

after 5 days at 24°C

tall healthy seedling

sickly yellow seedling

Figure 38.2   Intraspecific competition

The experiment shown in figure 38.2 is set up to investigate the effect of density on germinating cress seeds. Carton **A** with only 100 seeds represents low density of planting; carton **B** with 500 seeds represents high density.

After 7 days almost all the seeds in **A** are found to have successfully germinated and grown into healthy seedlings. Although most of the seeds in **B** also germinate, many fail to grow into healthy plants and remain yellow and sickly. Of the seedlings that do grow successfully, most are found to be smaller than their counterparts in **A**.

From this experiment it is concluded that the spaced out plants in **A** grow well because each plant receives an adequate supply of each growth requirement.

In carton **B** where the seedlings are densely packed, much of the light that might have reached each plant is intercepted by the leaves of other plants. This means that, on average, each plant receives less light for photosynthesis. In addition their rooting systems become interwoven and may be competing for water and minerals.

If a plant is short of water, its stomata stay closed for a longer time and carbon dioxide uptake (needed for photosynthesis) is reduced. Thus intraspecific competition between densely populated plants results in many individuals growing more slowly.

Commercial applications
Farmers use a drilling machine to space out crop seeds during planting. Horticulturalists mix tiny seeds with sand to sow them as thinly as possible and then thin out newly germinated seedlings to reduce intraspecific competition.

Interspecific competition

Plants of different species which occupy the same habitat often differ from one another in growth form (e.g. rooting depth, leaf shape etc.) and mineral requirements. Thus interspecific competition is normally less intense than intraspecific competition.

Nevertheless some species are able to become dominant at the expense of others. For example, the only plants able to survive amongst bracken are those able to grow and flower before the bracken fronds emerge. Pine trees almost totally prevent growth of any other plants on the forest floor by cutting out the light. Some plants are able to produce toxic substances which inhibit the growth of other species: the weed, gold of pleasure, makes a substance which suppresses the growth of nearby flax plants. This is of commercial importance since flax is cultivated for its fibre and seeds.

Effect of grazing on maintenance of species diversity

A completely unselective 'grazer' such as a lawn mower is found to maintain a **higher diversity** of plant species on a piece of grassland than occurs in its absence. Mowing keeps the more vigorous grasses in check. If they are left to grow normally, they compete with and gradually choke out and kill some of the less sturdy species.

Rabbits are relatively unselective grazers. If they are removed from a piece of grassland by the disease *myxomatosis* or by human activities, the aggressive dominant grasses are no longer held in check and they drive the other less vigorous species to extinction.

When the rabbits are present, they maintain a closely cropped area of grassland possessing a richly varied flora. This relationship is summarised in figure 38.3. At very high intensities of grazing, diversity of flora may be reduced slightly if the less vigorous species become overgrazed.

Although sheep are selective grazers, they also maintain species-rich communities since they select the competitive dominant grasses and hold them in check.

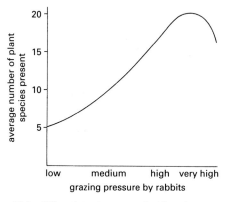

Figure 38.3 *Effect of grazing on species diversity*

Compensation point

At a certain low level of light intensity, the rate of photosynthesis occurring in a leaf exactly equals the rate of respiration. This particular light intensity is called the plant's **compensation point**. It varies from species to species.

At compensation point there is no net gaseous exchange. Neither is there a net gain nor loss of food. Figure 38.4 compares the events occurring in a leaf at compensation point with those occurring in darkness and in light of high intensity.

During 24 hours of clear weather, compensation point in a plant occurs twice: shortly after dawn and shortly before nightfall. During the day, between these two compensation points, photosynthesis exceeds respiration and a store of carbohydrate builds up. Some of this is needed by the plant for use during the night and on dull days.

Compensation period

This is the time taken by a plant to reach its compensation point after having been in darkness.

Sun and shade plants

Sun plants thrive in brightly illuminated habitats and grow best when they are not shaded by neighbouring members of the community.

Plants which thrive in dimly lit habitats where they are shaded for long periods of the year by other members of the community are called **shade plants**.

Consider the graph in figure 38.5 which compares a sun and shade plant with respect to their CO_2 output and uptake. Both plants give out CO_2 as a result of respiration in darkness. As the day dawns and light intensity gradually increases, each plant begins to photosynthesise

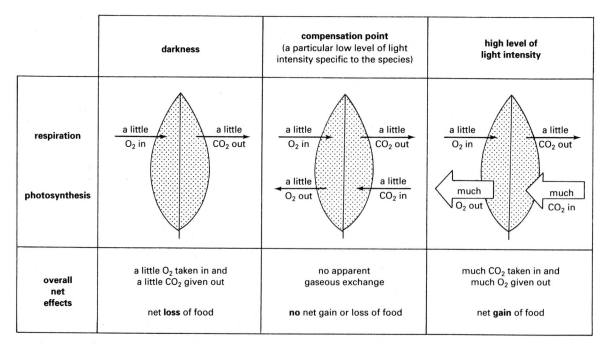

	darkness	compensation point (a particular low level of light intensity specific to the species)	high level of light intensity
respiration	a little O_2 in → a little CO_2 out	a little O_2 in → a little CO_2 out	a little O_2 in → a little CO_2 out
photosynthesis		a little O_2 out ← a little CO_2 in	much O_2 out ← much CO_2 in
overall net effects	a little O_2 taken in and a little CO_2 given out net **loss** of food	no apparent gaseous exchange **no** net gain or loss of food	much CO_2 taken in and much O_2 given out net **gain** of food

Figure 38.4 *Comparison of compensation point with two extremes*

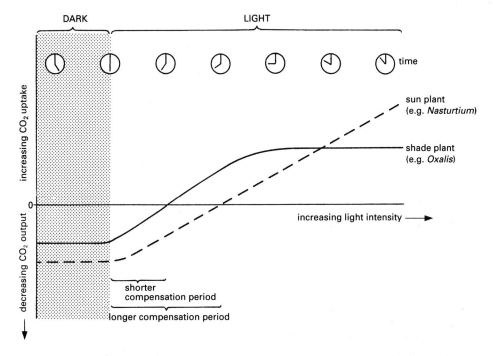

Figure 38.5 *Compensation periods of sun and shade plant*

until eventually photosynthetic rate equals and then exceeds respiration rate.

However, compared with the sun plant, the shade plant is found to have a *shorter* compensation period since its compensation point occurs at a *lower* light intensity.

Survival value
Since shade plants have a greater ability to use dim light, this adaptation enables them to survive in poorly illuminated habitats such as the floor of a deciduous forest. Such shade-tolerant plants produce their flowers and seeds early in the year before the dense leaf canopy above them blocks out most of the light and reduces their photosynthetic rate to a minimum.

number of grains planted per unit area	yield (grams of dry mass per unit area)
5	10
20	50
40	70
60	72
100	72

Table 38.1

cereal seed grains of the same species but differing in number from one another, were grown in areas of uniform size and soil fertility.
a) Plot the data as a line graph in the form of a curve.
b) (i) What relationship exists initially between number of grains planted and yield?
(ii) Why does this trend not continue indefinitely with increase in number of grains planted?
c) Apart from wastage of seed grain, what other disadvantage could result from planting too dense a crop?
d) Which of the following densities of planting would be likely to suffer most from interspecific competition:
A 20; **B** 40; **C** 60; **D** 100?
e) From your graph, state the optimum number of seed grains that should be planted per unit area.

QUESTIONS

1 Explain the meaning of the terms mobility and sessility with reference to living things.
2 a) State TWO environmental factors for which neighbouring plants could be competing.
  b) When an essential factor is in short supply, why is interspecific competition usually less intense than intraspecific competition?
3 Table 38.1 gives the results of a plant competition experiment where five groups of

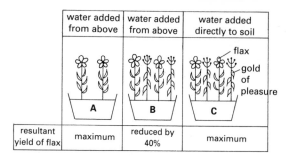

	water added from above	water added from above	water added directly to soil
	A	B	C
resultant yield of flax	maximum	reduced by 40%	maximum

Figure 38.6

4 The experiment shown in figure 38.6 as set up in an attempt to demonstrate that the weed, gold of pleasure, makes a chemical which inhibits growth of flax.

a) In which pot does inhibition appear to have occurred?

b) Suggest which part of the weed plant could be making the inhibitor. Explain your answer.

c) Why was control pot **A** included in the experiment?

d) State TWO ways in which the experiment would need to be improved before the experimenter would be justified in drawing any valid conclusions.

5 Two similar squares of turf (X and Y) were cut out of a piece of grassland and examined. Each was found to possess the same community of 20 different plant species. The squares were kept in a greenhouse under identical conditions except that X was regularly cropped whereas Y was left uncut.

After three years, one of the squares was found to possess 11 of the original species while the other still had all 20.

Identify which square was which and explain why.

6 **a** Define the terms compensation point and compensation period.

b) *'No plant can survive indefinitely in nature at the light intensity of its compensation point.'* Is this statement true or false? Justify your answer.

c) The graph in figure 38.7 refers to a single plant during a period of 24 hours in summer.

(i) For how long did the plant's morning compensation period last?

(ii) At what two times did the plant reach its compensation point?

(iii) By how many hours were these two points separated during the day?

(iv) Predict how your answer to (iii) would have differed if the graph had been drawn for the same plant in autumn.

EXPERIMENTAL DESIGN QUESTION

Observations: When light strikes the leaves of trees in a woodland community much of the red and blue light is absorbed by the leaf canopy. The light that reaches the ground flora of shade plants is therefore deficient in these wavelengths but still rich in yellow and green light.

Hypothesis: Compared with a related species of sun plant, a shade plant contains a richer supply of carotenoid pigments (in order to make efficient use of yellow light).

Problem: Design an experimental procedure which would enable you to test the validity of the above hypothesis. (See chapter 5 if you need help.)

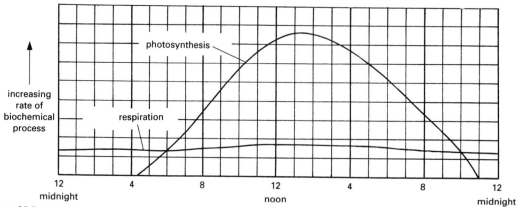

Figure 38.7

What you should know (CHAPTERS 37–38)

1 Many animals show distinct **behaviour patterns** when **foraging** for food.
2 Such behaviour tends to increase the animal's chance of gaining **maximum net energy**.
3 To be economical an animal must forage optimally with respect to **search and pursuit time**, **type of food** selected and **size of food** selected.
4 **Intraspecific** competition occurs between members of the **same** species. **Interspecific** competition occurs between members of **different** species.
5 Intraspecific competition for territories is called **territoriality**. It spaces out a population in relation to available food supply.
6 **Dominance hierarchy** amongst the members of a social group involves lower ranking individuals acknowledging the status of those with higher rank. They do this by showing **subordinate responses** to the latter's **threat displays**. This behaviour conserves energy and ensures experienced leadership.
7 **Co-operative hunting** benefits the members of a social group since all animals gain more food than they would hunting on their own.
8 Animals are mobile but plants are **sessile**. This normally poses no problem for plants since their immediate environment provides all the raw materials needed for survival.
9 **Intraspecific** competition exists amongst the members of a **dense** population of plants for **light**, **water** and **soil nutrients**.
10 **Interspecific** competition between plants tends to be **less intense** than between animals because different species often have different requirements.
11 **Grazing** of land by herbivores maintains **species diversity** since dominant plants are held in check.
12 **Compensation point** is that **low level of light intensity** at which rate of photosynthesis in a plant exactly equals rate of respiration.
13 **Compensation period** is the **time** taken by a plant which has been in darkness to reach its compensation point. It is shorter in shade plants than in sun plants.

39 Coping with dangers – animals

Avoidance behaviour and habituation

Snail

Consider the experiment shown in figure 39.1. When the underside of the glass sheet is tapped sharply, the snail shows **avoidance behaviour** by retreating into its shell. This escape **response** is an example of **unlearned** behaviour. Snails are genetically programmed to respond to danger in this way. (Such a behavioural adaptation is of survival value and has therefore been favoured by natural selection during millions of years of evolution.)

When the snail re-emerges and the glass is

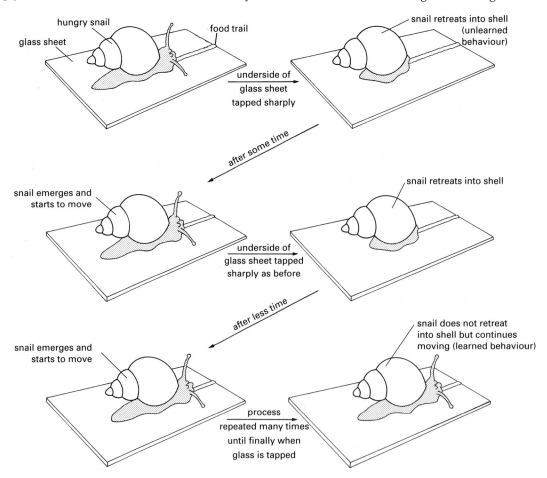

Figure 39.1   Avoidance behaviour and habituation

tapped for a second time, the snail retreats as before. However, after many repeats, the snail eventually ceases to respond to the stimulus and therefore fails to show avoidance behaviour.

This is because the stimulus has proved, after many repeats, to be **harmless** and the animal has learned *not* to react to it. This type of **learned** behaviour is called **habituation**. It is beneficial because it prevents the animal from performing its escape response so often that it has no time or energy left to carry out essential activities such as obtaining food.

After a period lacking stimulation, the animal does respond to the stimulus as before and does show avoidance behaviour. It is important that habituation is **short-lived** because a long-term modification of the escape response would leave the animal open to danger.

In this experiment, hungry snails are used so that they will be motivated to emerge from their shells and follow the food trail (flour in water). Many snails of the same species are used, environmental factors such as light and temperature are kept constant and the experiment is repeated many times.

Learning experiments

Mirror drawing

Using her normal writing hand, the learner (see figure 39.2) is timed as she joins up the dots forming the star outline while looking only at the mirror image.

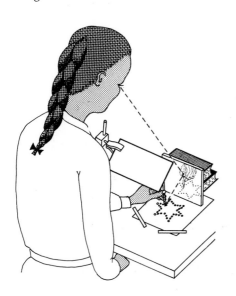

Figure 39.2   Mirror drawing

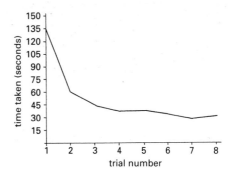

Figure 39.3   Learning curve (type 1)

After several repeats of the task, the time required to perform it successfully is found to decrease until a minimum is reached. When graphed the results give a **learning curve** (see figure 39.3).

Typing words

A non-typist is allowed one minute in which to correctly type as many three-lettered words as possible from a long list. When the procedure is repeated several times, the learner is found to increase the score. This is shown in figure 39.4 which is a second type of learning curve.

Long-term modification

When each learning experiment is repeated several days later, the learner is found to have retained some of the learning. Now the performance at trial 1 is better than it was during the first experiment.

This shows that learning involves a **long-term modification** of the response made to a stimulus. In order to learn, an animal must be capable of **remembering** a previous situation.

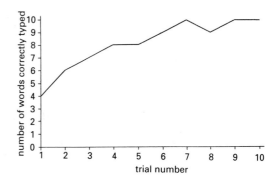

Figure 39.4   Learning curve (type 2)

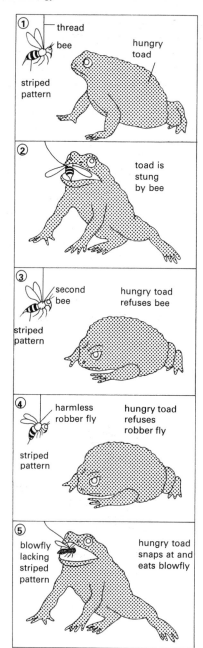

Figure 39.5 *Learning to avoid danger*

Learning to avoid danger

Toads

Toads feed on insects. A toad's natural instinct is to snap at any moving object that looks like an insect. The toad shown in figure 39.5 has previously been fed on harmless insects presented on the end of a thread but has never been presented with a bee.

When the toad rises to the bee bait, it receives a violent sting and spits the insect out. When offered a second bee, the toad ducks its head and refuses it. When offered a robber fly whose striped pattern closely resembles a bee, the toad refuses it but snaps at an insect that does not resemble the bee.

From this experiment it is concluded that the toad has learned that the yellow-striped pattern of the bee means danger and avoids it.

Birds

Newly hatched ducklings and goslings quickly learn to follow the first large object that they meet if it moves and gives out sounds. The young birds are responding to the visual and auditory stimuli normally provided by their mother whom they learn to follow.

This form of learning, which can only occur during a brief period of early life, is called **imprinting** (see figure 39.6). It is a behavioural adaptation of survival value. It provides the means by which a young bird avoids danger by staying close to and being protected by its mother.

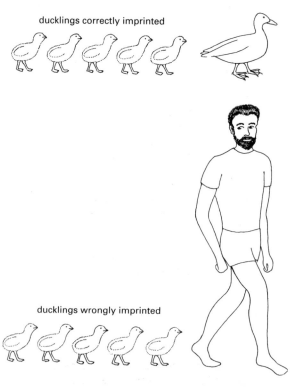

Figure 39.6 *Imprinting*

| Flammable | Toxic | Harmful | Biohazard |

| Explosive | Corrosive | Oxidising | Radioactive |

Figure 39.7   Hazard signs

Humans

Human beings learn to avoid danger by being educated by, and imitating, their more experienced and knowledgeable elders.

In a science laboratory, danger is avoided by learning the meanings of the **hazard signs** (see figure 39.7) and then applying appropriate precautions when using chemicals or materials bearing these signs.

Very young children are often unable to judge the speed and distance of approaching vehicles. Rather than learn 'road sense' by trial and error, they are taught the **Green Cross Code** and are accompanied when crossing the road.

Some learning by young children does however involve **trial and error**. Their power of curiosity sometimes overrides the warnings given by a more experienced adult. A child tempted, for example, to try touching nettle leaves soon finds out that they really do give a painful sting and learns to avoid them in future.

Intelligence
An animal's ability to learn depends on its intelligence. An intelligent animal is capable of **insight learning (reasoning)** and can solve a problem by applying previously learned concepts.

Advanced animals can therefore figure out what constitutes a potentially dangerous situation and take steps to avoid it. In recent years the installation and use of seat belts in cars has become legally enforced as humans have learned that this safety measure reduces the chance of fatal injury in a road accident.

Individual mechanisms of defence

Active

In addition to '**tooth and claw**', many other forms of active defence are found in the animal kingdom.

The skunk produces a **foul-smelling secretion** which it squirts at enemies. This causes great distress to the sprayed victim and allows the skunk to escape. Similarly many insects and reptiles defend themselves by injecting **poison** into their attackers. Some caterpillars are covered with **bristles** which cause considerable irritation to predators when touched.

Animals with long legs such as the antelope and ostrich defend themselves by **fleeing** from the enemy at top speed. Most birds take to the air when frightened.

Ground-nesting birds such as the avocet defend their young by performing **distraction displays**. Uttering plaintive cries, the adult lurches over the ground, helplessly flapping a 'broken' wing. Once the predator has been diverted from the nest, the adult suddenly flies off to safety.

A few animals attempt to defend themselves by **feigning death**. When threatened, the grass snake often turns on its back, opens its mouth, lets its tongue loll and plays dead until the enemy goes away.

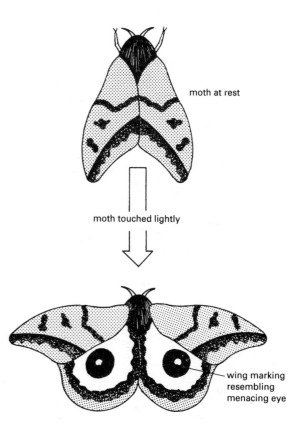

Figure 39.8   'Eye-spot' display in moths

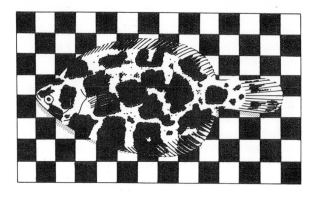

Figure 39.9 Camouflage

Passive

Many animals possess **protective coverings** (e.g. shell of limpet, armour plating of armadillo) which safeguard them from attacks by predators.

Markings on the wings of some moths (see figure 29.8) resemble menacing eyes which are effective at startling predators.

Camouflage is a further adaptation which provides prey animals with a means of passive defence. Some adult insects closely resemble leaves or twigs. Other harmless insects **mimic** the striped pattern of poisonous ones and are mistakenly ignored by predators. Some animals have **patterns** which help them to blend with their surroundings.

The flounder fish can even change the intensity of its colouring to resemble the background on which it is lying (see figure 39.9). This is a further example of an individual which possesses an inherited physiological adaptation, modifying its response to suit the needs of particular circumstances that it finds itself in (see also page 205).

Some animals (e.g. trout) possess a two-tone colour pattern called **counter shading** which offers double camouflage. The dark upper half of the body viewed from above is almost invisible against the river bed background while the light underside viewed from below blends effectively with the sky.

Social mechanisms for defence

By staying together as a large group (e.g. school of fish, herd of mammals), many types of animals rely on the principle of '**safety in numbers**' as a means of defence. Amongst a flock of birds, for example, there are many eyes constantly on the look out for enemies. Following an alarm, the bunching and swirling tactics adopted by the flock confuse the predator who finds it much more difficult to capture a member of a large unpredictable group than a solitary individual.

Defence is strengthened further by the members of a social group adopting a **specialised formation** as in the following examples.

Musk ox

Musk oxen (figure 39.10) are native to arctic regions of Canada and Greenland. Their natural enemy (apart from human beings) is the wolf. When threatened, a herd of musk oxen form a **protective group** with cows and calves in the centre. Individual wolves are gored and packs are driven off by a combined charge.

Figure 39.10 Social defence (musk ox)

Figure 39.11   Social defence (quail)

Quail

Bobwhite quails roost in **circles** with their heads to the outside as shown in figure 39.11. If disturbed, the circle acts as a defensive formation by 'exploding' in the predator's face. By the time their enemy has recovered from the confusion, the birds have flown away to safety.

Baboon

When a group of baboons is on the march (see figure 39.12), the dominant males stay in the centre close to the females with infants. Lower-ranking adult males and juveniles keep to the edge of the troop and raise the alarm if the group is threatened (e.g. by a leopard).

QUESTIONS

1 If an animal (e.g. a bird) is repeatedly exposed to a stimulus (e.g. a scarecrow) which fails to be dangerous, then eventually the animal fails to respond to the stimulus. Name this type of behaviour and explain why it is of advantage to the animal concerned.

2 *Nereis* is a type of marine worm which lives in a tube. Its head emerges from the tube to feed but is quickly withdrawn in response to a variety of stimuli.

   The graph in figure 39.13 shows the results of an investigation into the effects of two stimuli tried at two-minute intervals on 40 worms living in glass tubes in a shallow tank of water.

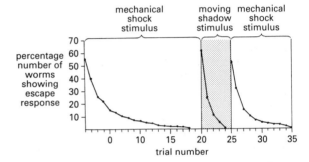

Figure 39.13

Figure 39.12   Social defence (baboon)

a) Why were as many as 40 worms used in the investigation?

b) Suggest how the mechanical shock stimulus was applied simultaneously to the worms.

c) Identify the unlearned escape response that *Nereis* employs to cope with such possible dangers.

d) (i) What evidence is there from the graph that the worms became habituated to both of the stimuli used in this investigation?

(ii) What evidence is there from the graph that one stimulus acts independently of the other?

e) (i) Which trial number's results best support the idea that habituation to mechanical shock is short-lived?

(ii) What is the survival value of this form of learned behaviour being short-lived?

3 During his first day in a new job, an employee was shown how to assemble a type of machine component. Figure 39.14 charts his progress during his first eight days of employment.

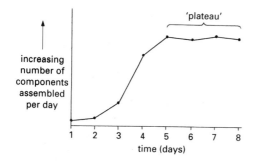

Figure 39.14

a) What name is given to this type of graph?

b) Suggest why the number of components assembled on day 2 was only slightly greater than on day 1.

c) Between which two days did 'learning' occur at the fastest rate? Suggest why.

d) Give ONE possible reason for the 'plateau'.

4 The upper half of a pigeon's body is darker than the lower half.

a) What name is given to this type of colour pattern?

b) Describe its dual role in camouflaging the bird.

c) Is this adaptation an individual or a social defence mechanism?

5 Amongst a herd of zebra, the stripes break up the animals' outlines making it difficult to tell where one animal ends and the next begins.

a) Suggest how this adaptation may help to protect the animals against a predator such as a lion.

b) Is such disruptive colouring an individual or a social mechanism for defence?

6 Some caterpillars depend on camouflage for defence whereas others rely on bright warning colours.

To be effective, which group should be clustered together and which group widely scattered? For each, explain why.

EXPERIMENTAL DESIGN QUESTION

Observations: The larva of the common gnat (*Culex*) is found to hang from the surface of calm water as shown in figure 39.15. When the pond water is disturbed slightly, the larva quickly swims downwards but soon returns to the surface.

In a tank in the laboratory, *Culex* larvae are found to remain at the surface despite its regular disturbance by bubbles from an aerator.

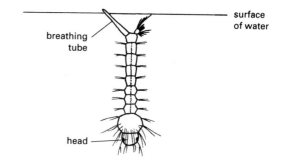

Figure 39.15

Hypothesis: The *Culex* larvae in the laboratory have become habituated to the water disturbance caused by the air bubbles.

Problem: Design an experimental investigation to test the validity of the above hypothesis.

DATA INTERPRETATION

The camouflage provided by cryptic coloration of body parts increases the chance of survival of many prey organisms. This affords little protection however to those organisms whose life style demands much mobility. Many such animals are found to have evolved an alternative means of protection: **mimicry**, the phenomenon of one species (especially an insect) closely resembling another.

Batesian mimicry involves an unequal partnership in which the mimic (an animal which is palatable to a predator) has come to closely resemble the model (an unpalatable, predator-resistant species). For example the conspicuous red and black coloration of the distasteful Monarch butterfly, which birds find especially noxious, is imitated by the harmless Viceroy butterfly.

Batesian mimicry is disadvantageous to predators since they are deceived into avoiding palatable prey or forced to run the risk of tackling distasteful/poisonous models by mistake. In addition, this relationship is also detrimental to the model whose security is undermined by any predator that eats a palatable mimic and is thereby encouraged to sample a similarly patterned model. Batesian mimicry only works, therefore, if the mimics are vastly outnumbered by the models; repeated stinging or poisoning of a predator by the models then renders the very rare encounter with a palatable mimic unprofitable.

Mullerian mimicry involves two or more predator-resistant species each of which has evolved a similar warning coloration. For example the yellow and black banded pattern shared by the stinging wasp and the distasteful cinnabar moth caterpillar. This form of mimicry is beneficial to both the predator (it only has to learn to avoid one warning pattern) and the prey (any losses suffered while the predator is still learning are shared by all the species involved).

a) To be successful why must cryptic coloration be accompanied by a high degree of immobility? (1)

b) From the passage, name TWO predator-resistant features other than coloration which are often possessed by prey animals. (1)

c) Batesian mimicry involves mimic, model and predator. Which of these is the sole beneficiary? In what way does this form of mimicry operate to the disadvantage of the other two? (3)

d) Batesian mimicry sometimes involves warning sounds instead of colouring. Explain how this would work. (1)

e) Why is mutual resemblance an advantage to all of the prey species concerned in a case of Mullerian mimicry? (1)

f) Which form of mimicry involves the members of one species acting as 'sheep in wolves' clothing' in order to deceive the members of another species? (1)

g) In addition to sharing a similar warning coloration, a group of distasteful unrelated species vulnerable to attack by nocturnal predators are all found to emit a similar odour. What sort of mimicry is this? Suggest how it operates. (2)

(10)

40 Coping with dangers – plants

Plant defences

Whereas animals can often learn to avoid danger and put up a fight if necessary or turn and flee, **sessile plants** must employ different strategies for defence.

Secondary compounds

Some plants are able to repel the attentions of herbivorous animals by producing **protective chemicals** in their tissues. These are called secondary compounds because they are by-products of the plant's basic metabolism. They act on the potential consumer in one of the following ways.

Chemical mimic

A tropical relative of the sedge family called grasshopper's cyperus makes a chemical which closely resembles juvenile growth hormone in insects. When consumed, this chemical disturbs the growing insect's hormonal balance resulting in the formation of an adult with twisted wings and, if female, underdeveloped ovaries.

Such insects are unable to lay eggs and cannot inflict future damage on plants. For this reason grasshopper's cyperus is being examined by genetic engineers. It would be ideal if the appropriate genes could be identified and introduced into crop plants.

The hairy wild potato plant mimics the alarm scent of aphids thereby keeping them at bay. For this reason this plant is also proving of interest to genetic engineers.

Some types of clover are able to make a substance which mimics the effect of oestrogen, a mammalian sex hormone. The herbivorous consumer (e.g. sheep) suffers reduced fertility (thereby cutting down future potential damage to the plants).

Poison

Most varieties of buttercup produce a chemical which is poisonous to farm animals. In addition the animals avoid meadow buttercup because of the acrid (bitter) taste of its leaves.

Cyanogenesis
Hydrogen cyanide is a poison which acts by blocking an organism's cytochrome system. Some varieties of white clover (*Trifolium repens*) contain non-toxic glycoside which is hydrolysed by enzyme action to hydrogen cyanide when leaf tissue is damaged (e.g. during nibbling by herbivores such as slugs). This production of cyanide is called **cyanogenesis**. Plants unable to make cyanide are said to be acyanogenic.

Table 40.1 shows the results from an experiment where several members of the same species of slug were released into plastic boxes each containing equal numbers of leaves from cyanogenic and acyanogenic clover plants. The

		% number of leaves		
		undamaged	less than half of leaf eaten	half or more of leaf eaten
type of clover leaf	cyanogenic	80(+)	16(−)	4(−)
	acyanogenic	42(−)	25(+)	33(+)

(+ and − indicate more or less than expected by chance)

Table 40.1 Damage to cyanogenic and acyanogenic clover leaves

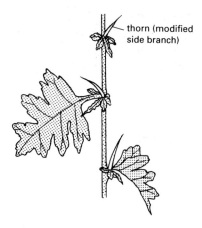

Figure 40.1 Thorn

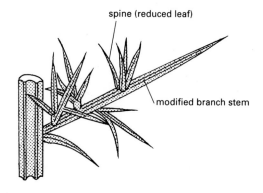

Figure 40.2 Spine

cyanogenic plants suffered much less damage.

Cyanogenesis does not always guarantee protection. For example, the larvae of the common blue butterfly produce a further enzyme which converts poisonous cyanide into harmless thiocyanate.

Structural defence mechanisms

Amongst flowering plants, a variety of **structural adaptations** are found, each designed to keep hungry herbivores at bay.

Thorns

Hawthorn trees bear sharp **thorns** which are modified side branches (see figure 40.1).

Spines

In gorse, the **leaves** are reduced entirely to short **spines** (figure 40.2) which are borne on modified branch stems. This presents an almost impenetrable barrier to hungry herbivorous mammals.

The leaves of some cacti (see page 213) are completely reduced to spines. This helps to protect inner succulent tissues from thirsty desert animals.

Investigating the number of spines on holly leaves
In holly, the spines are restricted to the edges of the leaves. Figure 40.3 shows the results of an

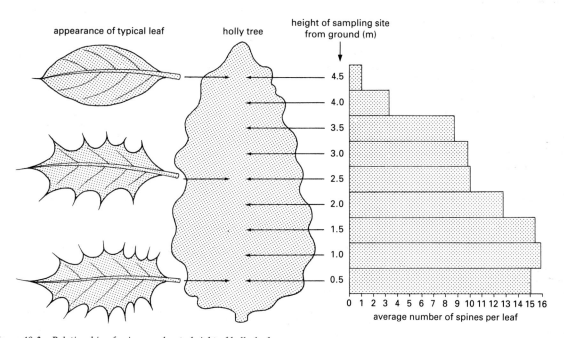

Figure 40.3 Relationship of spine number to height of holly leaf

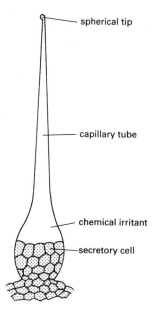

Figure 40.4   Sting

investigation into the relationship between number of spines present and height of leaf above the ground. 20 leaves were collected at each sample site; on average, the number of spines was found to decrease as the height of the leaf increased.

It is thought that natural selection has favoured those holly plants possessing spines concentrated on the lower leaves because this adaptation is of survival value. Leaves near the ground are within easy reach of many browsing herbivores and if they lacked spines they would be eaten. As height of leaf increases, the chance of attack is reduced and therefore the number of protective spines decreases.

Stings

The leaves of nettle possess **stinging hairs** (see figure 40.4). Each takes the form of a thin capillary tube ending in a spherical tip. When an animal touches a hair, its spherical tip breaks off leaving a sharp edge. This penetrates the animal's skin allowing liquid irritant to be injected, giving an impressive warning.

Ability to tolerate grazing

The growth form of many plants is specially adapted to survive continuous grazing. Most grasses, for example, have very **low growing points** and are therefore able to continue to send up new leaves despite the fact that their older

ones are constantly being eaten. This enables grass to withstand a degree of grazing that would be fatal to most other plants.

Many grasses also possess **rhizomes**. These underground storage organs are capable of asexual reproduction and help to ensure the plant's survival.

Plantain and dandelion have evolved a flat **rosette habit** (figure 40.5). Their leaves radiate out from a very short stem and often lie firmly pressed against the soil surface enabling them to escape the attention of grazing herbivores.

Some meadow plants also possess enormous powers of **regeneration** (see page 129). A small piece of dandelion root, for example, is capable of producing an entire plant.

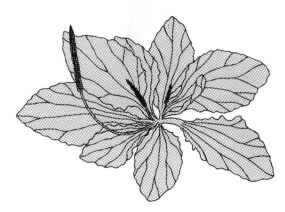

Figure 40.5   Rosette

QUESTIONS

1 **a)** What is cyanogenesis?
 b) How does it work as a defence mechanism in plants?
 c) Describe an entirely different type of defence mechanism employed by plants that is also based on the use of a secondary compound.
2 **a)** Present the information given in table 40.2 as a graph by plotting the points and then drawing the best curve.
 b) State the relationship that exists between cyanide concentration and age of leaf.
 c) In what way could this relationship be of survival value to the plant?
3 Describe TWO adaptations possessed by grass plants which enable them to tolerate heavy grazing.

age of bracken leaf (weeks)	average cyanide content of leaf (mg/100 g dry wt.)
1	9.9
3	4.3
5	2.1
7	1.0
9	0.5
11	0.2
13	0.0

Table 40.2

4 a) What relationship exists between number of spines on a holly leaf and distance of leaf from the ground?

b) Suggest why holly trees have evolved in this way.

c) Give ONE other structural defence mechanism employed by a named plant and briefly describe how it operates.

DATA INTERPRETATION

Cyanogenesis in *Trifolium repens* is controlled by two genes. One controls the presence (allele G) or absence (allele g) of glycoside in leaf cells. The other controls the presence (allele E) or absence (allele e) of the enzyme necessary to hydrolyse glycoside to hydrogen cyanide. This process only occurs when the leaf is damaged by, for example, a chewing herbivore or intense frost.

Figure 40.6 refers to the phenotypic frequencies of wild populations of *Trifolium repens* at different altitudes from the slopes of a mountain.

Table 40.3 shows the results of a series of experiments on the selective feeding habits of eight species of invertebrate animal offered only the two types of clover as food.

species of invertebrate herbivore	type of *Trifolium repens*	
	cyanogenic	acyanogenic
1	−	+
2	−	+
3	+	+
4	−	+
5	−	+
6	−	−
7	−	+
8	−	+

(+ = plant eaten   − = plant rejected)

Table 40.3

a) Would a plant with the genotype ggEe be cyanogenic or acyanogenic? (1)

b) Give all the possible genotypes that would result in a cyanogenic phenotype. (1)

c) (i) Construct a word equation to represent the process of cyanogenesis.

(ii) Why does a plant capable of cyanogenesis not poison itself to death? (1)

d) (i) What generalisation can be made from the results shown in the table? (1)

(ii) Which TWO species of herbivore are the 'exceptions to the rule'? (1)

(iii) What percentage of plants were able to make cyanide at altitude 600 m? (1)

(iv) Suggest a possible selective advantage enjoyed by such cyanogenic plants at such low altitudes. (1)

e) (i) What relationship exists between frequency of plants capable of cyanogenesis and increasing altitude? (1)

(ii) What percentage of plants at altitude 1900 m were unable to produce hydrogen cyanide? (1)

(iii) Suggest why such acyanogenic plants enjoy a selective advantage. (1)

(10)

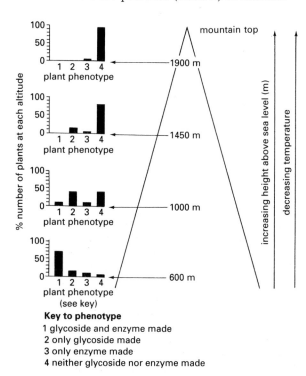

Figure 40.6

Key to phenotype
1 glycoside and enzyme made
2 only glycoside made
3 only enzyme made
4 neither glycoside nor enzyme made

What you should know (CHAPTERS 39–40)

1 Some animals learn *not* to react to a stimulus if, after many repeats, it proves to be harmless. This is called **habituation**.

2 Habituation prevents the animal **wasting time** and **energy** on needless repeats of its avoidance response.

3 Habituation only brings about a **short-term modification** of the escape response, otherwise the animal would be left open to danger.

4 In advanced animals, **learning** involves a **long-term modification** of the response made to a stimulus. Information is stored in the central nervous system and **remembered** for future use.

5 Animals show many **individual defence mechanisms** such as camouflage and protective coverings.

6 **Social mechanisms** for defence are found amongst animals which remain together in a protective group.

7 Some plants protect themselves using **chemicals**.

8 Some plants possess **structural** adaptations for defence such as thorns.

9 The **growth form** of some plants enables them to tolerate grazing animals.

Appendix 1 Ultrastructure of a cell

Magnification and resolution

Magnification is the apparent enlargement of an object. By magnifying material, a microscope allows structures which are invisible to the naked eye to be examined.

Resolution (resolving power) is the ability of a microscope to produce a separate image of each of two structures located closely together.

When two objects lie more closely together than the limit of resolution of the miscroscope, they appear as a single fused image. Further magnification simply increases the size of the fused image.

The maximum useful magnification possible using a **light microscope** is approximately 1500 times. An **electron microscope** has a much higher resolving power and can achieve a useful magnification of over 500 000 times.

Electron micrographs

An image of material viewed under an electron microscope can be recorded as a black and white

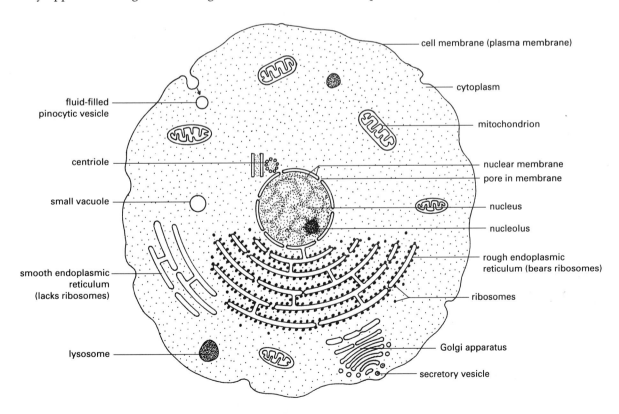

Figure A1.1 Ultrastructure of generalised animal cell

photograph called an **electron micrograph**. Figures A1.1 and A1.2, which are based on several electron micrographs, show how the electron micrsocope reveals the presence in a cell of many tiny structures which cannot be seen using a light microscope. Many of these specialised structures are called **organelles** and they make up the cell's **ultrastructure**.

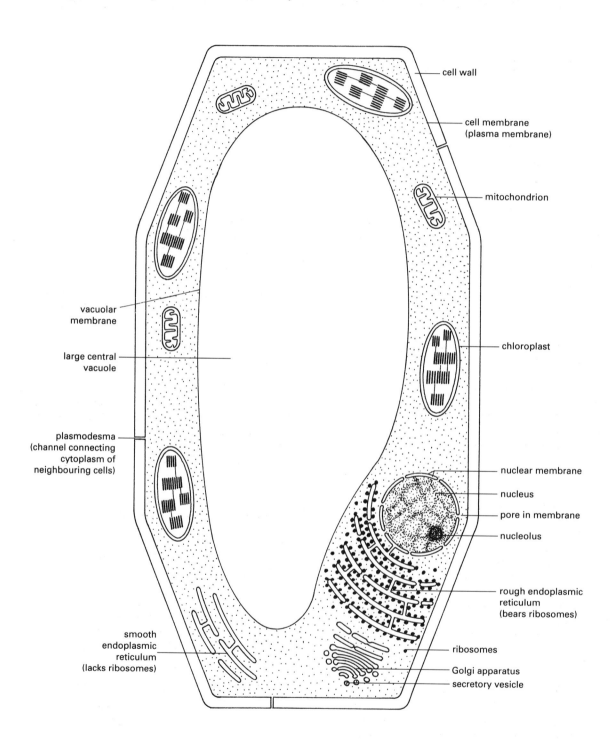

Figure A1.2   Ultrastructure of generalised plant cell

Appendix 2 Water potential

Water potential (Ψ) is the potential of a cell (or solution) to **give out water** by osmosis. (The pressure at which this water moves through a selectively permeable membrane is measured in units called pascals.)

Osmotic potential (**OP**) is the potential of a cell (or solution) to **take in water** by osmosis and exert an osmotic pressure (OP is also measured in pascals.) OP is determined by the concentration of solute molecules present in the solution. The more solute present, the higher the OP.

Water potential of pure water and aqueous solutions

Ψ and OP are equal and opposite. Thus $\Psi = -\text{OP}$.

Since pure water contains no solutes, OP of pure water = 0 and Ψ of pure water = 0.

Since movement of water molecules is hindered by the presence of neighbouring solute molecules, pure water has the highest possible Ψ.

In other words, *zero is the highest water potential possible* and all solutions have lower water potentials which are expressed in negative values. The more concentrated the solute content of a solution, the more negative its Ψ.

Water always moves from high Ψ (less negative) to low Ψ (more negative). If two cells or solutions have the same Ψ, they are **isotonic**. If they differ in Ψ, then the one with the higher Ψ is said to be **hypotonic** to the one with the lower Ψ which is **hypertonic**.

Isolated cell in bathing solution

Consider figure A2.1.

When $\Psi c > \Psi e$, water passes from the cell to the bathing solution and the cell may become **plasmolysed**.

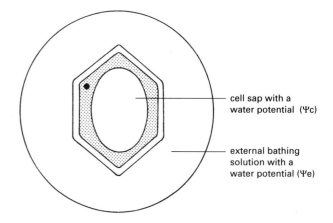

cell sap with a water potential (Ψc)

external bathing solution with a water potential (Ψe)

Figure A2.1 Isolated cell

When $\Psi c = \Psi e$, there is no net gain or loss of water by the cell or the bathing solution and the cell remains **unchanged**.

When $\Psi c < \Psi e$, water enters the cell from the bathing solution and the cell gains **turgor**. (See also chapter 2.)

Neighbouring cells in a tissue

Consider figure A2.2.

When Ψ cell X > Ψ cell Y, water passes from cell X to cell Y.

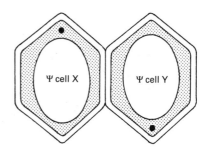

Ψ cell X Ψ cell Y

Figure A2.2 Neighbouring cells

When Ψ cell X = Ψ cell Y, there is no net flow of water.

When Ψ cell X < Ψ cell Y, water passes from cell Y to cell X.

Water potential gradient

Normally the soil solution surrounding a root has a higher water potential than the contents of root cells. Water therefore enters root hairs by osmosis resulting in these cells possessing a higher Ψ than adjacent cortex cells. Water therefore passes from the root hairs into the cortex cells and so on across the root to the xylem vessels along a **water potential gradient** (see figure A2.3).

Similarly water passes from xylem vessels to leaf cells along a Ψ gradient. (See also chapter 36).

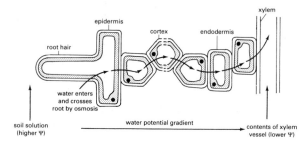

Figure A2.3   Water potential gradient

Water potential of fish

A difference in Ψ exists between the internal and external environments of both fresh water and salt water bony fish as summarised in figure A2.4. The maintenance of water balance in these animals is described in chapter 35.

Summary

Table A2.1 summarises the terms that apply to two cells (or solutions) separated by a selectively permeable membrane where cell (or solution) **X** has a higher water potential than cell (or solution) **Y**.

cell (or solution) X	cell (or solution) Y
higher water potential (ψ)	lower water potential (ψ)
higher water concentration	lower water concentration
lower osmotic potential (and pressure)	higher osmotic potential (and pressure)
lower solute concentration	higher solute concentration
hypotonic	hypertonic
loses water by osmosis	gains water by osmosis

Table A.2.1   Summary

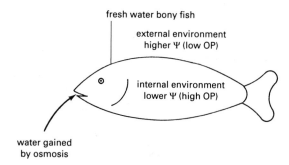

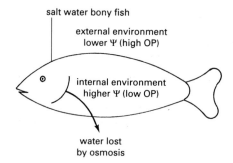

Figure A2.4   Water potential of fish

Appendix 3 Designing an experimental procedure to test a hypothesis

Guidelines

Living organism under investigation

Remember to read the hypothesis carefully and identify the living organism (or biological material) to which it specifically refers.

Prediction

To test a hypothesis you begin by making a **prediction**. This involves anticipating in advance some **change or difference** that the organism (or material) could show which would support the hypothesis.

This change or difference could, for example, be:

- change in weight;
- difference in growth rate;
- presence or absence of a product;
- difference in population density etc.

When you make your prediction ensure that it indicates the measurements (or observations) that would need to be made in order to establish that the change had occurred (or difference was evident).

'Golden' rules

Read table A3.1 to revise the **'golden' rules** that apply to biological investigations. Then decide how each of these rules would apply to this particular investigation (making rough notes if necessary).

Control

Consider whether your predicted results (if obtained) would be sufficient on their own to support the hypothesis or if you would also need to include a **control** for purposes of comparison.

'golden' rule	reason
only one variable may be altered at a time in an investigation	if more than one variable were altered then it would be impossible to know which one was responsible for the results obtained
a relatively large amount of living material is used (e.g. many organisms/cells/samples etc.)	if only a relatively small amount of living material were used then it might be unusual and atypical of the species in general
the results of the experiment can be successfully repeated	if the results cannot be repeated then they may have been the outcome of a lucky chance which does not constitute valid support for the hypothesis

Table A.3.1   The 'golden' rules

Precautions and safety

Identify any **special precautions** that would need to be included in your design (e.g. checking of airtight seals, careful use of flammable chemicals etc.).

Experimental procedure

With all of these points in mind (and aided by your rough notes) select the ones that are relevant to this particular investigation and give a **step by step** account of the procedure that you would adopt to test the hypothesis.

Example of experimental design question

Untreated cells from a fresh beetroot were viewed under the microscope. Cell samples from the same beetroot, which had been immersed in one of three different concentrations of sucrose solution, were then examined.

Observations: Compared with untreated cells, those in 0.15 M (molar) sucrose appeared more turgid, those in 0.35 M sucrose appeared to be plasmolysed but those in 0.25 M sucrose seemed to be unchanged.

Hypothesis: Beetroot cells gain water when immersed in 0.15 M sucrose solution, lose water in 0.35 M sucrose but neither gain nor lose water in 0.25 M sucrose.

Problem: Design an experimental procedure that would enable you to test the validity of the above hypothesis.

Answer

Living organism under investigation

Fresh live beetroot tissue.

Prediction

A sample of beetroot tissue immersed in 0.15 M sucrose solution will increase in weight, one in 0.35 M sucrose will decrease in weight but one in 0.25 M sucrose will neither increase nor decrease in weight.

Experimental procedure

1 Using a cork borer, cut out several cylinders of tissue from a fresh beetroot.
2 Slice these into many thin discs of equal size.
3 Blot the discs dry and divide them into three equal groups.
4 Weigh each group accurately (e.g. using an electronic balance) and record the results.
5 Immerse a group of discs in a beaker of 0.15 M sucrose solution.
6 Repeat step 5 using equal volumes of 0.25 M and 0.35 M sucrose solutions.
7 Place the beakers in a thermostatically controlled water bath at 25° C to maintain uniform temperature throughout the experiment.
8 After one hour remove each group of discs, dry them and reweigh.
9 Express any increase or decrease in weight as a percentage.
10 Find out if similar results have been obtained by other groups performing the same investigation (using different beetroots).
11 Check if the pooled results match the original prediction in order to find out whether or not the hypothesis is supported.

Appendix 4 The genetic code

		second letter of triplet				
		A	G	T	C	
first letter of triplet	A	AAA AAG AAT AAC	AGA AGG AGT AGC	ATA ATG ATT ATC	ACA ACG ACT ACC	A G T C
	G	GAA GAG GAT GAC	GGA GGG GGT GGC	GTA GTG GTT GTC	GCA GCG GCT GCC	A G T C
	T	TAA TAG TAT TAC	TGA TGG TGT TGC	TTA TTG TTT TTC	TCA TCG TCT TCC	A G T C
	C	CAA CAG CAT CAC	CGA CGG CGT CGC	CTA CTG CTT CTC	CCA CCG CCT CCC	A G T C

(A = adenine, G = guanine, T = thymine, C = cytosine)

Table A4.1 DNA's bases grouped into 64 (4 × 4 × 4) triplets

codon	anti-codon	amino acid	codon	anti-codon	amino acid	codon	anti-codon	amino acid	codon	anti-codon	amino acid
UUU UUC	AAA AAG	} phe	UCU UCC	AGA AGG	} ser	UAU UAC	AUA AUG	} tyr	UGU UGC	ACA ACG	} cys
UUA UUG	AAU AAC	} leu	UCA UCG	AGU AGC		UAA UAG	AUU AUC	} •	UGA UGG	ACU ACC	• tryp
CUU CUC CUA CUG	GAA GAG GAU GAC	} leu	CCU CCC CCA CCG	GGA GGG GGU GGC	} pro	CAU CAC CAA CAG	GUA GUG GUU GUC	} his } glun	CGU CGC CGA CGG	GCA GCG GCU GCC	} arg
AUU AUC AUA	UAA UAG UAU	} ileu	ACU ACC ACA ACG	UGA UGG UGU UGC	} thr	AAU AAC AAA AAG	UUA UUG UUU UUC	} aspn } lys	AGU AGC AGA AGG	UCA UCG UCU UCC	} ser } arg
AUG	UAC	met									
GUU GUC GUA GUG	CAA CAG CAU CAC	} val	GCU GCC GCA GCG	CGA CGG CGU CGC	} ala	GAU GAC GAA GAG	CUA CUG CUU CUC	} asp } glu	GGU GGC GGA GGG	CCA CCG CCU CCC	} gly

(U = uracil)

Table A4.2 in RNA's 64 codons, tRNA's 64 anticodons and amino acids coded

abbreviation	amino acid
ala	alanine
arg	arginine
asp	aspartic acid
aspn	asparagine
cys	cysteine
glu	glutamic acid
glun	glutamine
gly	glycine
his	histidine
ileu	isoleucine
leu	leucine
lys	lysine
met	methionine
phe	phenylalanine
pro	proline
ser	serine
thr	threonine
tryp	tryptophan
tyr	tyrosine
val	valine
●	chain terminator

Table A4.3   Key to amino acids

Appendix 5

Chi-squared test

Since breeding experiments in genetics involve an element of **chance**, observed results tend to **deviate** from the expected results. In order to decide if such a deviation is significant or not, use is made of a statistical procedure called the 'goodness of fit' or **chi-squared** (χ^2) **test**.

Once calculated, the χ^2 value expresses the deviation as a single numerical value. By locating the position of this number in a table of distribution values for χ^2, the probability (p) of obtaining such a deviation can be found.

If this probability is 5 per cent or more, then the deviation is regarded as being the result of chance alone (and not a real departure from the expected results).

Calculating χ^2

The formula for calculating χ^2 is:

$$\chi^2 = \Sigma \frac{(O-E)^2}{E}$$

where Σ = sum of, O = observed results and E = expected results.

Worked example

In the experiment on page 65, Mendel's observed F_2 results were 5474 round and 1850 wrinkled pea seeds out of a total of 7324. Based on a 3:1 ratio, the expected results would be 5493 round and 1831 wrinkled.

Each phenotype is said to form a class.

Thus $\chi^2 = \dfrac{(O-E)^2}{E}$ round $+ \dfrac{(O-E)^2}{E}$ wrinkled

When this is calculated as shown in table A5.1, χ^2 is found to be 0.263.

Reference is now made to the table of the distribution values for χ^2 (see table A5.2). The degrees of freedom referred to in the table allow for the number of independent comparisons involved in the test. This is always one less than the number of classes being considered. When the row for two classes and one degree of

class	O	E	O−E	(O−E)²	$\dfrac{(O-E)^2}{E}$	χ^2
round	5474	5493	−19	361	$^{361}/_{5493} = 0.066$	0.066 + 0.197 = 0.263
wrinkled	1850	1831	+19	361	$^{361}/_{1831} = 0.197$	

Table A5.1 Calculating χ^2

number of classes	degrees of freedom		increase ← **probability (p) values** → decrease									
			0.90 (90%)	0.80 (80%)	0.70 (70%)	0.50 (50%)	0.30 (30%)	0.20 (20%)	0.10 (10%)	0.05 (5%)	0.02 (2%)	0.01 (1%)
2	1	$\chi^2 \longrightarrow$	0.016	0.064	0.148	0.455	1.074	1.642	2.706	3.841	5.412	6.635

↑ 0.264 lies here
← deviation not significant → ← deviation significant →

Table A5.2 Distribution values for χ^2

253

freedom is examined, a value of $\chi^2 = 0.263$ is found to fall somewhere between 0.148 (probability 70%) and 0.455 (probability 50%). This probability is well above the accepted minimum of 5% showing that, in this case, the deviation is not significant and is simply due to chance. The observed results are therefore accepted as showing a 3:1 ratio.

Index

Answers

James Torrance

Diagrams by James Torrance

Answers

1 Cell variety in relation to function

1 A red blood cell's small size and biconcave shape allow it to present a relatively large surface area of cell to the surrounding environment. It also contains haemoglobin which has an affinity for oxygen. These features exactly suit it to its function of uptake and transport of oxygen.

2 **a)** Water transport and mechanical support.
   b) It takes the form of a hollow tube strengthened by rings or spirals of lignin.

3 **a)** See figure A1.1.

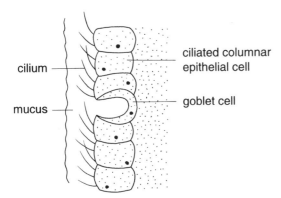

Figure A1.1

   b) A goblet cell is cup-shaped and secretes mucus which traps dirt. A columnar epithelial cell has hair-like cilia which sweep dirty mucus up and away from the lungs.

4 **a)** X is a root hair and it absorbs water and mineral salts.
   b) Y consists of sieve tubes and companion cells. Each sieve tube has sieve plates but no nucleus whereas each companion cell lacks sieve plates but has a nucleus. A sieve tube's function is to transport sugars; a companion cell's function is to control a sieve tube.
   c) Yes. A root is arranged into different tissues with each type specialised to perform one (or more) particular function(s).

5 **(i)** They both possess columnar epithelial cells and goblet cells.
   (ii) Unlike the windpipe lining, the epithelium from the intestine has enzyme-producing cells and lacks cilia.

6 **a)** Sperm.
   b) It has a tail for swimming towards an egg. Its rich supply of mitochondria provides plenty of energy for active movement.

7 **a)** By photosynthesis.
   b) **(i)** By using its photoreceptor and eye spot.
   (ii) By whipping its flagellum to propel itself along.
   c) It lacks the wall typical of plant cells yet it has chloroplasts which animal cells do not possess.

2 Absorption and secretion of materials

1 Diffusion, osmosis and active transport.

2 **a)** It is a non-living layer consisting of cellulose fibres and pectic substances enclosing water-filled spaces.
   b) **(i)** Fluid mosaic model.
   (ii) **A** = phospholipid; **B** = protein.
   (iii) **C** and **D**.
   (iv) It provides the means by which small molecules can pass through the membrane.

3 **a)** **(i)** Oxygen.
   (ii) Carbon dioxide.
   b) Protein (or starch).

4 **a)** X = diffusion; Y = active transport.
   b) Y.   **c)** X.
   d) It will decrease. It requires energy from respiration but respiratory enzymes do not work well at low temperatures.

5 **a)** Hypotonic. Since the ratio for potato cells in 0.2M sucrose is less than 1, their final mass must have been greater than their initial mass, showing that water passed into the cells from the sucrose solution by osmosis.
   b) 0.26M.

6 A, E, C, B, F, D.

Experimental design answer

See Appendix 3 of text book.

Data interpretation answers

a) **(i)** Ion concentration is greater in the cell sap than in the pond water.
   (ii) Potassium = 1200; sodium = 71.
   (iii) The data supports the selective theory since it shows that the plant accumulates different concentrations of different ions (rather than accumulating equal concentrations of them all).

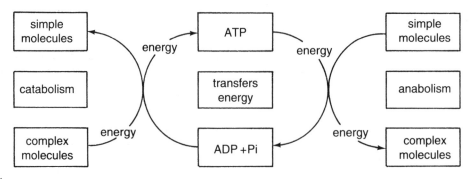

Figure A3.1

(iv) It disputes this suggestion. In every case given in this example, ion uptake is from low to high concentration which is the reverse of diffusion.

b) The plant makes more carbohydrate by photosynthesis in bright light than in dim light. More energy is therefore available in bright light for active uptake of ions.

c) (i) It results in an increase in rate of ion uptake.
(ii) Some other factor (e.g. amount of sugar available for energy release) has become limiting.
(iii) It is an inverse relationship. Since sugar provides the energy for ion uptake, the number of units of sugar present in the cell sap decreases as the number of units of ion absorbed increases.

d) (i) Approximately 37°C.
(ii) Respiratory enzymes are denatured at higher temperatures.

e) Following drainage, the soil became warmer and, as air entered the spaces between the soil particles, oxygen reached the root hairs. Therefore increased active uptake of essential ions occurred leading to improved growth of the crop plants.

3 ATP and energy release

1 a) Adenosine triphosphate.
b) ADP + Pi + energy → ATP.
c) It is able to make energy available for energy-requiring processes.

2 a) and **b)** See figure A3.1.

3 A rapid turnover of ATP molecules occurs constantly in the body's cells.

4 They would be unable to grow and increase in number. They would fail to make additional ATP since this process would be limited by the shortage of inorganic phosphate.

5 Oxidation involves the removal of hydrogen from the substrate and the release of energy. Reduction involves the addition of hydrogen to the substrate and the consumption of energy.

4 Chemistry of respiration

1 Its inner membrane is folded into cristae which present a large surface area on which chemical processes can take place.

2 a) See table below.

process	site where process occurs in cell	reaction(s) involved in process	products
A glycolysis	*[cytoplasm]*	splitting of *[glucose]* into 3-carbon compound	ATP and *[pyruvic acid]*
B *[Krebs]* cycle	central matrix of *[mitochondrion]*	removal of *[hydrogen]* from carbon compounds by oxidation; release of carbon atoms forming *[carbon dioxide]*	carbon dioxide
C *[oxidative phosphorylation]* (hydrogen acceptor system)	*[cristae]* of mitochondrion	release of energy from hydrogen (oxidation of hydrogen)	*[ATP]* and *[water]*

b) Krebs cycle.

c) Krebs cycle and oxidative phosphorylation.

3 For each molecule of glucose, aerobic respiration produces 38 ATP, carbon dioxide and water, whereas anaerobic respiration produces 2 ATP, carbon dioxide and ethanol.

4 The concentrations of lactic acid and carbon dioxide increase. Since both of these are acidic, they depress the pH.

5 **a)** The clip that is open should be closed and an amount of sodium hydroxide equal to that in tube **Y** should be put into the bottom of tube **X**.

b) By using a thermostatically controlled water bath.

c) **A** will move down; **B** will move up. Carbon dioxide given out by the worm is absorbed by the sodium hydroxide. Oxygen taken in by the worm causes a decrease in volume of the enclosed gas. Level **B** rises to fill this space causing level **A** to drop.

d) Their movement will be much slower. Respiration will proceed at a reduced rate since respiratory enzymes will be affected by the low temperature. (Unlike a human, a worm cannot maintain a high body temperature in cold conditions.)

e) To keep the experiment fair and prevent the introduction of a further variable factor.

f) To invalidate the argument that the results were simply due to some space inside the respirometer being occupied and not due to the worm respiring.

g) Before beginning the experiment, a syringe would be fitted to tube **Y** with its needle through the stopper. After the experiment had been running for a known length of time, the syringe would be used to find out the volume of air needed to return **B** to its original level. This would give a reading of the worm's oxygen consumption/unit time.

5 Absorption of light by photosynthetic pigments

1 **a)** It means that some light passes through the leaf.

b) Absorption and reflection.

2 **a)** Blue and red.

b) Most photosynthesis occurs at these regions on the strand of alga. These sites therefore release most oxygen which in turn attracts many aerobic bacteria.

3 Chlorophyll b appears green since this is the colour of light which it does not absorb but gives

back out again. Phycoerythrin appears red since this is the colour of light which it fails to absorb and gives back out again.

4 An absorption spectrum shows which wavelengths of light have been absorbed by leaf pigments. An action spectrum shows which wavelengths of light have been used for the process of photosynthesis.

5 **a)** Grana are coin-like stacks of sacs containing chlorophyll; stroma is a colourless background material lacking chlorophyll.

b) The grana are the site of the light-dependent stage; the stroma is the site of the carbon fixation stage.

Experimental design answer

Living organisms under investigation
Red seaweed and dandelion.

Prediction
A chromatogram of photosynthetic pigments from the red seaweed will differ from that of dandelion leaves by one or more spots.

Experimental procedure
1 Grind up equal masses of both plants in sand and acetone.
2 Filter to give two pigment extracts.
3 Repeatedly spot and dry each extract on to separate strips of chromatography paper.
4 Add chromatography solvent to two boiling tubes and replace the stoppers.
5 Run the two chromatograms by allowing the solvent to ascend the papers.
6 Pool results of other groups and decide which give the most typical result for each separation.
7 Compare two typical chromatograms and find out if they match the prediction and therefore lend support to the hypothesis (or not).

Data interpretation answers

a) **P** = chlorophyll a.
A = chlorophyll c, xanthophyll and carotene.
b) Fungus.
c) See figure A5.1. (overleaf).
d) See x-axis of graph in figure A5.1.
e) (i) **Pigment 2** = phycocyanin.
(ii) **Pigment 1** = phycoerythrin.
f) See figure A5.2. (overleaf).
g) Since phycoerythrin absorbs green light, this extends the range of wavelengths that can be absorbed (beyond those taken in by chlorophyll a) and helps the plant to survive in a dimly lit habitat.

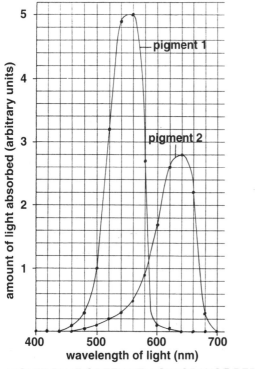

VIOLET BLUE GREEN YELLOW ORANGE RED

Figure A5.1

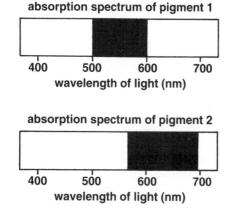

absorption spectrum of pigment 1

absorption spectrum of pigment 2

Figure A5.2

6 Chemistry of photosynthesis

1 **a)** It is split into hydrogen and water.
   b) It is the regeneration of ATP from ADP and inorganic phosphate using energy from light absorbed by chlorophyll.
2 **a)** Glycerate phosphate and ribulose bisphosphate.
   b) GP becomes converted to RuBP in light.
   c) In light, both substances remain at a steady level but in darkness the amount of GP rises and that of RuBP falls. This provides evidence that the conversion occurs in light (with a steady state being maintained as the cycle turns) but that the conversion stops in darkness (with GP accumulating and RuBP disappearing as the cycle becomes disrupted).
3 **a)** See figure A6.1.
   b) Reduced hydrogen acceptor and ATP.
   c) Carbon dioxide.
4 **a)** As bubbles of oxygen released/unit time or as carbon dioxide uptake/unit time.
   b) Light intensity.
   c) Carbon dioxide concentration.
   d) 20.
   e) Temperature.

7 DNA and its replication

1 **a)** Sugar (deoxyribose) and phosphate.
   b) See figure A7.1.

Figure A7.1

2 See figure A7.2.

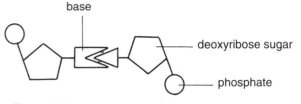

Figure A7.2

3 30%.
4 DNA, the four types of DNA nucleotide, enzymes and ATP.

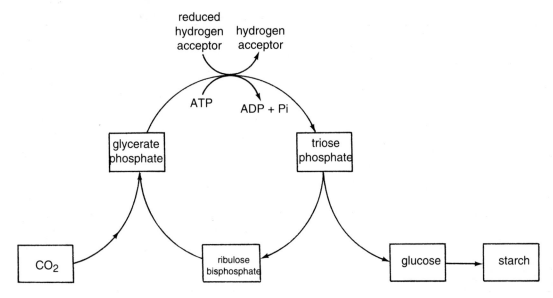

Figure A6.1

5 c, a, d, b, f, e.

6 50%.

7 a) A human red blood cell lacks a nucleus.
b) That of a chicken does have a nucleus but that of a cow does not.
c) For each type of animal, the DNA content of its sperm is half that of its liver/kidney cells.
d) DNA content is at its highest following DNA replication during interphase. It returns to normal when the growing cell undergoes mitosis and its nucleus divides into two.

8 RNA and protein synthesis

1 See figure A8.1.

Figure A8.1

2 DNA is double-stranded and possesses deoxyribose sugar and the base thymine; RNA is single-stranded and possesses ribose sugar and the base uracil.

3 a) Asparagine = AAU; glutamic acid = GAA; proline = CCU; threonine = ACC; tyrosine = UAU.
b) See figure A8.2.

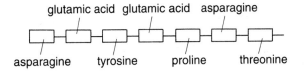

Figure A8.2

4

stage of synthesis	site
formation of mRNA	nucleus
collection of amino acid by tRNA	cytoplasm
formation of codon-anticodon links	ribosome (in cytoplasm)

5 a) tRNA collects another molecule of its particular amino acid.
b) mRNA is used to make another molecule of the same protein.
c) The completed protein is transported via the endoplasmic reticulum to the Golgi apparatus to be processed.

6 This is because protein structure is determined by the base sequence present in DNA which varies from person to person.

7 a) The increase in relative numbers of ribosomes in cells during days 1 to 5 indicates that this is the period of most rapid protein synthesis and growth of the new leaf. After day 5 growth slows down and comes to a halt at day 11 when the leaf has reached its full size.

b) A basic number of ribosomes will always be needed by a cell of a fully grown leaf to make proteins such as enzymes essential for biochemical pathways such as photosynthesis.

8 **a)** **A** = cell membrane
B = vesicle
C = Golgi apparatus
D = pore
E = nuclear membrane
F = chromosome
G = ribosome
H = endoplasmic reticulum.
b) Protein.
c) **G**.
d) An organism's genetic code consists of a particular sequence of bases present in its DNA. Each amino acid is coded for by a triplet of these bases. The sequence in which amino acids occur in a protein is therefore originally determined by the order of the bases in DNA. (Information is transferred from DNA to protein during transcription and translation.)

9 Variety of proteins

1 Nitrogen.
2 They can become arranged in long parallel strands or they can become folded together into a spherical shape.
3 Structural proteins which form an essential part of all cellular membranes; enzymes which speed up the rate of biochemical processes; and antibodies which defend the body against antigens.
4 See figure A9.4.

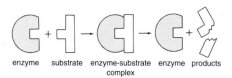

enzyme   substrate   enzyme-substrate   enzyme   products
complex

Figure A9.4

5 **a)** Casein contains them all. Group 2 rats gained weight throughout the experiment. Zein lacks two essential amino acids. Group 1 rats lost weight throughout the experiment.
b) (i) Zein.
(ii) Their diet could have been changed to casein or to zein supplemented with the two essential amino acids that it lacks.
c) 20%.
d) 1.5.

Data interpretation answers

a) Failure to transport oxygen to the body tissues.
b) The variation in their electrical charge (some are neutral, some positive and some negative).
c) (i) A suitable solvent is allowed to equilibrate with the air in the tank. The paper is inserted into the solvent keeping the origins above the solvent level. The solvent front is allowed to rise to the top of the paper.
(ii) The variation in their degree of solubility in the solvent.
d) (i) Neutral.
(ii) Positive.
Being neutral, peptide X has remained in the centre; being positively charged, peptide X^1 has moved to the left, towards the negative side, during electrophoresis.
e) It means the working out of the correct order in which amino acids appear in a peptide.
f) A molecule of glutamic acid in haemoglobin A is replaced by a molecule of valine in haemoglobin S.

10 Viruses

1 Nucleic acid and protein coat.
2 (i) Inside a host cell.
(ii) In the absence of suitable host cells.
3 **C, F, B, G, D, A, E**.
4 **a)** They are able to multiply.
b) It lacks sub-cellular structures such as ribosomes, nucleus, mitochondria and cell membrane.
5 **a)** DNA, helical, enveloped.
b) (i) **X** = human immunodeficiency virus (HIV), **Y** = tobacco mosaic virus.
(ii) Reverse transcriptase.
c) Herpes.
d) 1 RNA   . . . go to **2**
DNA   . . . go to **5**
2 helical   . . . go to **3**
polyhedral . . . go to **4**
3 naked   **tobacco mosaic virus**
enveloped **influenza virus**
4 naked   **bushy stunt virus**
enveloped **human immunodeficiency virus**
5 helical   . . . go to **6**
polyhedral . . . go to **7**
6 naked   **coliphage virus**
enveloped **smallpox virus**
7 naked   **bacteriophage virus**
enveloped **herpes**

11 Cellular response in defence

1 First line: blood clotting and lysozyme production.
Second line: antibody production and phagocytosis.
2 **d, b, c, a.**
3 The enzymes are kept enclosed inside lysosomes and vacuoles.
4 **a)** Lymphocyte.
   b) See figure A11.1.

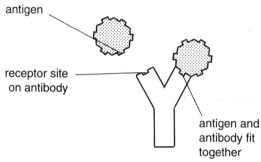

antigen

receptor site on antibody

antigen and antibody fit together

Figure A11.1

5 **a)** In response to the antigen (even in a damaged state) lymphocytes make antibodies which give immunity. In addition some of these lymphocytes act as memory cells allowing a rapid response to the antigen if invasion occurs in the future.
   b) Artificially.
6 **a)** To suppress the immune system in order to prevent rejection of the transplanted kidney.
   b) Susceptibility to diseases such as pneumonia.
7 The plant produces antifungal chemical called phytoalexin which prevents the invader spreading beyond the site of attack.
8 **a)** Willow tree.
   b) (i) Larva.
   (ii) Many uninfected young willow trees would be grown in a greenhouse. These would be divided into four groups. Group 1 would be injected with chemical extract from the eggs of the insect; group 2 with chemical from the larvae; group 3 with chemical from the pupae; and group 4 with chemical from the adults. If galls developed on the willow plants in group 2 only, this would support the suggestion.

Data interpretation answers

a) (i) IgG.

(ii) The baby's own immune system makes antibodies after a few months.
b) (i) IgM.
   (ii) IgE.
c) 1.4–4.0 mg/ml.
d) The concentration of IgM increases before that of IgG in both the primary and the secondary response.
e) During the primary response, the concentration of IgG takes a longer time to begin rising than during the secondary response. The highest concentration of IgG reached during the primary response is less than that reached during the secondary response. The concentration of IgG decreases after reaching its maximum level in the primary response but remains level at its highest concentration in the secondary response.
f) Since the response to the second injection was faster than to the first, this suggests that certain lymphocytes already 'knew what to do' from the previous time.

12 Meiosis

1 Haploid refers to a cell containing a single set of chromosomes; diploid refers to a cell containing a double set of chromosomes which form pairs.
2 Mitosis occurs in a plant's meristems and all over a growing animal's body, whereas meiosis only occurs in sex organs.
   Homologous chromosomes form pairs during meiosis but not during mitosis.
   Crossing over may occur during meiosis but never during mitosis.
   Mitosis involves one nuclear division whereas meiosis involves two nuclear divisions.
   Mitosis results in the formation of two identical daughter cells whereas meiosis results in four genetically different gametes.
   Chromosome number remains unaltered following mitosis but is halved by meiosis.
   Variation within a population is increased by meiosis but not by mitosis.
3 **a)** 12.
   b) 6.
   c) 6.
   d) 3.
   e) 3.
   f) (i) See figure A12.1. (overleaf).
   (ii) See figure A12.2. (overleaf).
4 **a)** 2.
   b) See figure A12.3. (overleaf).

arrangement 1

arrangement 2

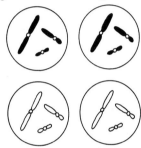

Figure A12.1

gametes from arrangement 1

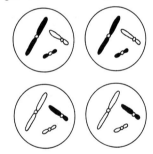

gametes from arrangement 2

Figure A12.2

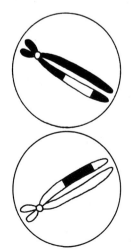

Figure A12.3

13 Monohybrid cross

1 a)

cross	ss × Ss
gametes	all s ↓ S and s
F₁	ss and Ss

b) 50%.

2 a) 6.

b) W.

c) (ii) 1 in 2.

3 The rabbit of unknown genotype would be crossed with a rabbit possessing uniform coat. If the unknown rabbit were homozygous, the following results would be obtained:

cross	SS × ss	
gametes	all S ↓ all s	
F₁	all Ss	(all spotted)

If the unknown rabbit were heterozygous, the following results would be obtained:

cross	Ss × ss
gametes	S and s ↓ all s
F₁	Ss and ss (1 spotted: 1 uniform)

4 a) and b)

(i)

cross	RR × RR	
gametes	all R ↓ all R	
F₁	all RR (all red, no ratio)	

(ii)

cross	RW × RW	
gametes	R and W ↓ R and W	
F₁	RR, RW, RW, WW	
	(1 red: 2 roan: 1 white)	

(iii)

cross	RW × WW
gametes	R and W ↓ all W
F₁	RW and WW
	(1 roan: 1 white)

(iv)

cross	RR × WW
gametes	all R ↓ all W
F₁	all RW
	(all roan, no ratio)

5 Cross 1 = $CC^H \times C^aC^a$.
Cross 2 = $C^{ch}C^a \times C^HC^a$.

14 Dihybrid cross

1 a) and b) See figure A14.1.

2 a) eevv.

b) ev.

c) EV, eV, Ev and ev.

d) Second parent = EeVv, B = Eevv, C = eeVv and D = EeVv.

3 a) Genes R, S and T which have their loci on the same chromosome, are said to be *linked* genes.

Let round = R and pear shape = r
Let red = C and yellow = c

cross 1 RRCC × rrcc

gametes all RC ↓ all rc

F₁ all RrCc

cross 2 RrCc × RrCc

gametes RC,Rc,rC,rc ↓ RC,Rc,rC,rc

male gametes

	RC	Rc	rC	rc
RC	RRCC	RRCc	RrCC	RrCc
RC	RRcC	RRcc	RrcC	Rrcc
rC	rRCC	rRCc	rrCC	rrCc
rc	rRcC	rRcc	rrcC	rrcc

(female gametes — left vertical label)

F₂ phenotypic ratio = 9 round red:
3 pear-shaped red:
3 round yellow:
1 pear-shaped yellow:

Figure A14.1

b) Crossing over, the process which can separate the alleles of such genes, occurs at points called *chiasmata*.
c) There is a greater chance of crossing over occurring between genes *S* and *T* than between genes *R* and *S*.
d) If a cross-over occurred between loci S and T then the recombinant gametes formed would be *rSt* and *RsT*, and the parental gametes would be *rST* and *Rst*.

4 **a)** and **b)**
cross 1 wwff × WWFF
gametes all wf ↓ all WF
F₁ all WwFf
cross 2 WwFf × wwff
gametes WF, Wf, wF, wf ↓ all wf
F₂ WwFf, Wwff, wwFf, wwff
81 14 16 89
white, white, brown, brown,
full shrunk full shrunk
(R) (R)
(R = recombinant)

c) (i) 15%.
(ii) Because the two genes are on the same chromosome and show linkage.
d) 15 units.

Experimental design answer

Living organism under investigation
True-breeding strains of grey-bodied, straight-winged fruit flies and yellow-bodied, curved-winged fruit flies.

Prediction
A cross between the above two strains of fruit flies will give an F₁ generation which on being self-fertilised will produce an F₂ with the phenotypic ratio 9 grey-bodied, straight-winged: 3 grey-bodied, curved-winged: 3 yellow-bodied, straight-winged: 1 yellow-bodied, curved-winged.

Experimental procedure
1 Obtain male grey-bodied, straight-winged and virgin female yellow-bodied, curved-winged fruit flies (or vice versa).
2 Using ether, anaesthetise both strains of flies and put a few males and females into several culture tubes.
3 Incubate the tubes at a suitable warm temperature and remove the parents after the eggs have been laid.
4 Once the F₁ adults emerge, repeat steps **2** and **3** using F₁ males and females.
5 When the F₂ generation emerge, anaesthetise them and classify them according to their phenotypes.
6 Find out (preferably using statistical analysis) if the results are close enough to 9:3:3:1 to agree with the prediction and therefore support the hypothesis (or not).

Data interpretation answers

a) It is based on the fact that neither of them is expressed in the phenotype of the F₁.
b) Normal wing, purple eye and vestigial wing, normal eye.
c) Crossing over (at chiasmata).
d) (i) 18.5%.
(ii) 17.5%.
(iii) It returns the grouping of the alleles to the parental combination.
e) (i) See figure A14.2.

Figure A14.2

(ii) It is not possible to say on which side of the other gene the new one is situated if only one recombination value is given.

f) (i) It must be situated immediately next to P/p.

(ii) WwBb × wwbb (or WwVv × wwvv).

15 Sex linkage

1 original cross $\qquad$ $X^rY \times X^RX^R$

gametes $\qquad$ X^r and $Y \downarrow$ all X^R

F_1 $\qquad$ X^RX^r and X^RY

second cross $\qquad$ $X^RX^r \times X^RY$

gametes $\qquad$ X^R and $X^r \downarrow X^R$ and Y

F_2 $\qquad$ $\underbrace{X^RX^R, X^RX^r,}\ X^RY,\ X^rY$

$$2 \quad : \quad 1 \quad : \quad 1$$

red-eyed, red-eyed, white-
females male eyed,
male

2 See figure A15.1

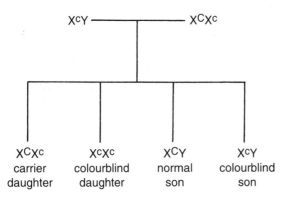

Figure A15.1

3 a) $A = X^HY$, $B = X^HX^H$, $C = X^hY$, $D = X^HX^h$.

b) Male parent = X^HY and is normal. Female parent is X^HX^h and is normal (but a carrier).

c) None.

4 a) It is impossible for a male to be tortoise-shell. Since the **Y** chromosome carries no allele for coat colour, the heterozygous genotype cannot arise in a male.

b)

possible cross 1 $\qquad$ $X^BX^B \times X^GY$

$\downarrow$

offspring $\qquad$ X^BX^G (and X^BY)

possible cross 2 $\qquad$ $X^GX^G \times X^BY$

$\downarrow$

offspring $\qquad$ X^GX^B (and X^GY)

5 a) All the F_1 offspring would be white. A cross involving a non sex-linked gene would give the same results.

b)

cross $\qquad$ $X^wX^w \times X^WY$

gametes $\qquad$ all $X^w \downarrow X^W$ and Y

F_1 $\qquad$ X^Wx^w and X^wY

$$1 \quad : \quad 1$$

white, red,
male female

A ratio of 1 white male: 1 red female in the F_1 is different from the all white F_1 that would result from a cross involving a non sex-linked gene. Such a 1 : 1 ratio for both sex and phenotype would therefore verify that the gene for plumage colour is sex-linked.

16 Mutation

1 A mutation is a change in an organism's genetic material such as the inheritance of an extra copy of chromosome 21 in humans. A mutant is an organism (e.g. human) who has inherited the altered genetic material which in turn affects her/his phenotype such as a sufferer of Down's syndrome.

2a) Substitution.

b) Inversion.

3a) Cell 1 = deletion; cell 2 = duplication.

b) Deletion.

4a) 2n = 18.

b) It possesses two different genomes whose members are unable to form homologous pairs.

c) Complete non-disjunction.

5a)

scientific name	common name	diploid chromo- some number (2n)	haploid chromo- some number (n)
Brassica oleracea	cabbage	18	9
Brassica rapa	turnip	20	10
Brassica rapobrassica	swede	38	19

b) (i) 2n = 19.

(ii) It possesses two different genomes whose members cannot form homologous pairs.

(iii) Complete non-disjunction could have taken place in a meristematic cell undergoing mitosis. This could have formed a polyploid cell (2n = 38) which grew into swede.

(iv) Soak root tips of the hybrid in colchicine to induce complete non-disjunction. Tissue-culture cells from the meristems of the affected roots. If these grow into plants, find out if their cells have the chromosome complement 2n = 38 and if their flowers can make viable gametes.

6 The graph shows increasing incidence of Down's syndrome with increase in maternal age. Since non-disjunction of chromosome pair 21 is the cause of Down's syndrome, the graph supports the given statement.

7 500 per million.

8a) (iii) aa × aa.
  b) 1 in 71429 chance.

9a) Mustard gas and X-rays.
  b) Mutations affecting sex cells can be transmitted indefinitely from generation to generation. Mutations affecting body cells only affect one generation.

10 It is possible that radiation somehow affects the genetic material in the gamete mother cells of some of the workers in nuclear processing plants. Perhaps this makes the genes which control blood formation undergo mutation at a higher rate than the normal spontaneous rate. If so then there would be a greater chance of these workers passing on the defective genes which cause leukemia to their children.

Data interpretation answers

a) In AB, the chromosomes in genome A cannot pair with those in B and so normal meiosis fails to occur. In AABB, the members of genome A pair with the members of the other A and likewise B with B, so gametes are formed.
b) The cross between them produced a sterile hybrid.
c) (i) Fertilisation of a gamete of one genome by one from a different genome (producing a sterile hybrid).
   (ii) Complete failure of spindle fibres during mitosis (leading to polyploidy).
   Allopolyploidy. *Triticum vulgare* contains three different types of genome.
d) 7.
e) *Triticum Vulgare* × rye
$$(2n = 42) \downarrow (2n = 14)$$
$$\text{sterile hybrid}$$
$$(2n = 28)$$
$$\downarrow$$
$$Tricale$$
$$(2n = 56)$$

17 Natural selection

1 Despite *over-production* of offspring, a population

explosion does not occur because the environment does not provide enough food. *Competition* for this (and other) limited resources occurs. Since *variation* exists amongst the members of the population, some will be better adapted (e.g. faster, stronger, etc.) and be able to survive. The weaker animals will lose out in the struggle and die.

2 **a)** The black one.
  b) See figure A17.1. The ratio = 1:1.

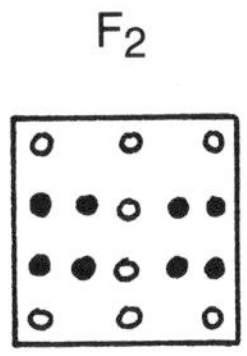

F₂

Figure A17.1

  c) See figure A17.2.

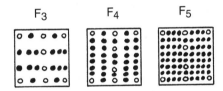

F₃   F₄   F₅

Figure A17.2

  (i) F₃.
  (ii) F₅.

3 Strains of bacteria resistant to the antibiotics will enjoy a selective advantage and could eventually replace the non-resistant wild type strains under this system. If any of the resistant strains turned out to be disease-causing, they would be very difficult to deal with.

4 The sickle cell allele (in the heterozygote) gives resistance to malaria and therefore enjoys a selective advantage in Africa where malaria is common. The incidence of the sickle cell trait is very low in non-malarial America where it is not favoured by natural selection.

5 **a)** An inverse relationship. The percentage number of melanic moths decreases as the number of lichen species increases.
  b) Melanic (dark) moths lose their selective advantage in clean places outside the city where they are no longer camouflaged and therefore easily spotted by predators.
  c) The line graph for pale moths would have shown a trend similar to that of the lichens.
  d) Polluted areas are becoming cleaner in

response to the Clean Air Acts. As a result the melanic moths are losing their selective advantage.

6 a) (i) Early dog violet.
(ii) Heath dog violet.
b) (i) Heath dog violet.
(ii) Neither one.
c) (i) Damp, neutral.
(ii) Highly competitive sturdy grasses keep them out.

Data interpretation answers

a) Consumption of warfarin interferes with the blood-clotting mechanism and the rat bleeds to death if it becomes cut or injured.
b) It thins their blood and prevents the formation of unwanted internal clots.
c) Mutation.
d) Strictly speaking it is incompletely dominant. Heterozygote W^sW^r which is resistant yet only needs some extra vitamin K in its diet is phenotypically intermediate to the two homozygous forms.
e) It has been favoured by natural selection since members of the sensitive strain normally die after eating the warfarin.
f) Extinction.
g) Very few W^rW^r will be fortunate enough to find a continuous supply of food containing the large amount of vitamin K that they need.
h) (i) $W^sW^r \times W^sW^r$
$$\downarrow$$
W^sW^s, W^sW^r, W^sW^r, W^rW^r
(ii) Yes. Since W^sW^r has the selective advantage (not needing unrealistic amounts of vitamin K yet still being resistant to warfarin), W^s will still keep occurring and surviving in the heterozygotes.

18 Selective breeding of animals and plants

1 The F_1 generation shows hybrid vigour.
2 a) They could cross them in the hope of producing a hybrid with both the desirable features.
b) The desirable features might be controlled by recessive alleles and therefore fail to be expressed in the phenotype of the hybrid.
3 a) It is the process by which breeders select individuals with certain desirable characteristics and use them to breed the next generation. They prevent the others from breeding.
b) Artificial selection enables the breeder to

accumulate those parts of a species' genotype which are useful to humans in a particular strain of animal or plant. The latter is then mass produced for the benefit of humans.

4 No. Eventually all the alleles for the desirable features will have been accumulated and the strain will not improve any further.

19 Loss of genetic diversity

1 If it possesses several desirable characteristics, a crop of uniform plants is of great use to humans. However the crop could eventually suffer inbreeding depression making it economically inferior.
2 a) Inbreeding leads to increased homozygosity. Homozygosity occurs even faster as a result of repeated selfing (e.g. self-pollination) compared with brother-sister mating.
b) It reduces the chance of two heterozygous carriers of some recessive inferior characteristic producing offspring homozygous for the inferior characteristic.
3 Agree. When the population is small there is a greater chance that all of the individuals with a certain allele will die if disaster strikes. That allele will therefore be lost.
4 a) Humans will continue to need other species for food, fuel and medicines.
b) On-site protection means maintaining an ecosystem containing a valuable gene pool in its natural state, whereas off-site protection means maintaining potentially valuable species in an artificial environment.
c) It is a storehouse of seeds and sex cells. It allows the alleles of thousands of wild varieties of crop plants (especially those that might become extinct) to be conserved for future use.

Data interpretation answers

a) F_1.
b) F_1 is taller, has longer mean ear length and has higher mean yield than either parent.
c) P_1.
d) Inbreeding depression means the decline of a certain characteristic as a result of continuous selfing as shown by plant height in this example.
e) (i) The yield has increased dramatically and less land is needed.
(ii) By doing so they are guaranteed a bumper F_1 crop whereas grains kept back by themselves would give poorer and poorer plants.

20 Genetic engineering

1 **a)** Dominant. The heterozygote has bar eyes.
  b) Duplication.
  c) 16.
2 Endonuclease cuts open DNA and ligase seals two ends of DNA together.
3 The gene for somatotrophin was located on its human chromosome. The gene was cut out using endonuclease and sealed into an opened plasmid from a bacterium using ligase. The recombinant plasmid was inserted into a host bacterial cell which was propagated on a large scale to produce somatotrophin. (This is then given to potential pituitary dwarfs during childhood.)
4 It may become possible to insert the appropriate genes into crops. The plants would then produce useful substances such as blood-clotting agent.
5 Somatic fusion.

21 Speciation

1 No because offspring resulting from the cross are not fertile.
2 It is possible that as the Earth became colder, all eggs were incubated at a lower temperature which produced one but not both sexes. This made reproduction impossible.
3 See figure A21.1.
4 **a)** (iv), (i), (iii), (v), (ii).
  b) Geographical, ecological and reproductive.
5 **a)** Some of the mutations unique to its isolated gene pool have been selected making it different from the mainland species.
  b) Continued isolation for a very long time.
6 **a)** They had evolved elsewhere and reached Australia by boat.
  b) They had evolved elsewhere and sailed or flew to New Zealand.

Data interpretation answers

a) Quantitative = body length (or tail length). Qualitative = ventral colour (or presence or absence of pectoral spot).
b) They are able to interbreed and produce fertile offspring.
c) (i) The genetic distance between this type of mouse and the one on Norway is much less than that between this mouse and the one on the Scottish mainland suggesting that the ancestor was Norse.
  (ii) It lends support to Berry. After the Ice Age,

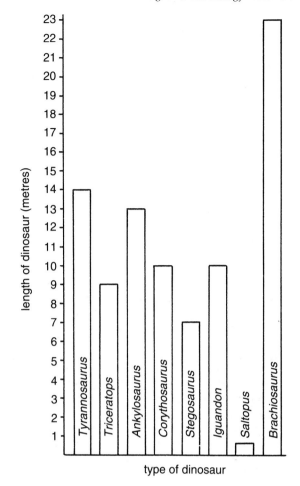

Figure A21.1

when *Apodemus* returned to the Scottish mainland (probably from southern parts of UK), a barrier of water separated the mainland from the Shetlands. So the route from Norway seems more likely.
d) If *A. s. fridariensis* on Fair Isle did arise from *A. s. granti* from Yell then the difference in their genetic distances can be explained as follows. The colonisers from Yell comprised a small splinter group which were genetically atypical of *A. s. granti* as a whole. This group eventually gave rise to a population with some gene frequencies very different from their Yell ancestors (i.e. genetic drift).
e) A common ancestor could have spread throughout the islands along with the human inhabitants in olden times when conditions were primitive. Different groups of mice have been kept isolated by water barriers ever since and have started to take different courses of

evolution in isolation (but not enough time has elapsed yet for complete speciation to have occurred).

22 Adaptive radiation

1 a) (i) 18 cubic units.
(ii) Desert dweller = 78 square units.
Eskimo = 72 square units.
(iii) The desert dweller's long slim body shape presents a larger surface area from which excess heat can be lost, thus keeping her/him cool in a hot climate. The Eskimo's short dumpy body shape presents a smaller surface area from which heat can be lost, thus helping to conserve heat and keep her/him warm in a cold climate.
b) (i) Adaptive radiation means the evolution of a group of related organisms along different lines by adapting to different environments. In the human example each group has taken its own course of evolution from a common ancestor. Natural selection has tended to favour the tallest slimmest people in the hot climate and the shortest most dumpy in the cold one.
(ii) They are no longer separated by effective barriers.
2 a) Analogous. They are only superficially alike and do not share a common ancestor.
b) Convergent.
3 They do not share a closely related common ancestor. They have arisen independently and then, by coincidence, have taken similar courses of evolution by becoming adapted to suit equivalent ecological niches on two separate continents.

Data interpretation answers

a) 3.
b) Isabella (or Española).
c) Santa Cruz.
d) (i) The percentage number of species only found on a particular island increases as the island's location becomes more distant from the centre of the group.
(ii) Isabella and Santa Fe have a similar percentage number of finches peculiar to them yet they vary enormously in size.
(iii) The more isolated islands have had less exchange with other gene pools and so a relatively higher number of their finches have taken their own separate course of evolution.
e) X = large seed-eating from Pinta.
Y = large cactus-eating from Pinta.
Z = medium seed-eating from Culpepper.

f) (i) Type X (large seed-eating).
(ii) One type might become extinct (or the two species might survive as small populations living in fierce competition or . . .?).

23 Growth differences between plants and animals

1 a) Growth is the irreversible increase in dry mass of an organism (normally accompanied by increase in cell number).
b) Regeneration is the replacement of lost or damaged parts by an organism.
2 a) It is a group of undifferentiated plant cells capable of mitosis.
b) Root tip and shoot tip.
c) Production of new cells.
3 a) Cell elongation and vacuolation.
b) (i) See figure A23.1.

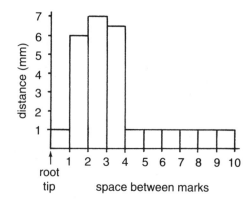

Figure A23.1

(ii) Cell elongation occurs between marks 1 and 4. Cells behind mark 4 have reached their full length.
4 A meristematic cell is alive, undifferentiated and capable of mitosis whereas a xylem vessel is dead, differentiated and incapable of mitosis.
5 A ring of cambium.
6 a) 6 years
b) Medullary ray. Its function is lateral transport of water and mineral salts.
c) Year 4.
d) Year 5. The fact that the ring is very narrow indicates a year of poor growth. This could have been due to decreased food production by damaged leaves.
7 a) Cambium.

b) B consists of spring wood whose vessels are bigger and wider than those of the autumn wood found at **A**.
c) Grain of timber.
8 The climate hardly changes from one season to the next so growth continues at a fairly constant rate throughout the year.

24 Growth patterns

1 a) See figure A24.1.

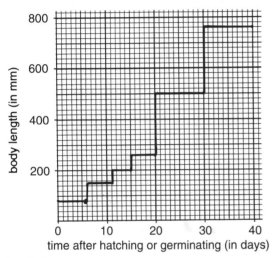

Graph 1

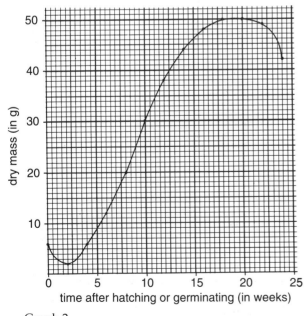

Graph 2

Figure A24.1

b) Graph 1's step-like form is typical of an insect's growth curve with abrupt increases in body length following moults. Graph 2's sigmoid form with a dip at the start is typical of an annual plant's growth with food reserves being used up at the start during seed germination.
2 Growth occurs only at meristems in a rose bush whereas it occurs all over a rabbit's body. A rose bush continues to increase in size throughout its life whereas a rabbit stops at adulthood. Regenerative powers are extensive in a rose bush but limited in a rabbit.

25 Genetic control

1 a) The appropriate gene is switched on by a light-sensitive reaction.
b) Chloroplasts would be of no use to the cell and their formation would be a waste of resources.

2

cross	$M^1m^1M^2m^2 \times m^1m^1m^2m^2$			
gametes	$M^1M^2, M^1m^2, m^1M^2, m^1m^2$ ↓ all m^1m^2			
off-spring	$M^1m^1M^2$ m^2	$M^1m^1m^2$ m^2	$m^1m^1M^2$ m^2	$m^1m^1m^2$ m^2
	brown	light brown	light brown	white

3 a)

	$B^1b^2B^3$	$B^1b^2b^3$	$b^1b^2B^3$	$b^1b^2b^3$
$b^1B^2b^3$	✓	×	×	×
$b^1b^2b^3$	×	×	×	×

Thus 1 in 8 chance.

b)

	$B^1b^2b^3$	$b^1b^2b^3$
$b^1B^2b^3$	×	×
$b^1b^2b^3$	×	×

Thus 0 chance.

c)

	$B^1b^2B^3$	$B^1b^2b^3$	$b^1b^2B^3$	$b^1b^2b^3$
$B^1B^2b^3$	✓	✓	✓	×
$B^1b^2b^3$	✓	×	×	×
$b^1B^2b^3$	✓	×	×	×
$b^1b^2b^3$	×	×	×	×

Thus 5 in 16 chance.

26 Control of gene action

1 a) (i) The regulator gene produces the *repressor* molecule.
(ii) The inducer molecule combines with the *repressor*.
(iii) When the operator is free, the structural gene is switched *on*.
b) (i) The operon that codes for this enzyme only becomes free to do so when lactose, the inducer, is present to combine with the repressor.
(ii) It prevents energy and resources being wasted on the production of an enzyme whose substrate might be unavailable.
c) No β-galactosidase will be produced since transcription and translation of the correct message will be impossible.
2 a) A mutated gene cannot code for its enzyme so the pathway becomes blocked.
b) (i) **Q**.
(ii) **R**.
(iii) **T**.
3 a) Excess phenylalanine is converted to tyrosine.
b) The enzyme needed to promote this conversion is absent.
c) It is converted to other metabolites.
4 a) See figure A26.1.
b) (i) Gonadotrophic = follicle-stimulating hormone (FSH). Ovarian = oestrogen.
(ii) Oestrogen.
(iii) A molecule of hormone combines with a receptor on the membrane of a target cell. The complex formed enters the cell. The hormone enters the nucleus. One or more genes become switched on and code for the appropriate proteins (e.g. keratin).

27 Hormonal influences on growth – part 1

1 a) Somatotrophin.
b) They are transported from one site to another site (the target tissue) where they bring about an effect.
c) Acromegaly occurs during adulthood and its effect (increased rate of growth) is restricted to certain parts of the body such as hands, feet and jaws. Pituitary dwarfism occurs during adolescence and its effect (reduced rate of growth) occurs all over the body.
2 F, E, B, D, A, C.
3 a) Downwards. It will probably respond to the

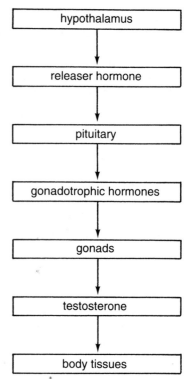

Figure A26.1

stimulus of gravity and show positive geotropism.
b) Bright light from above is a second variable factor. The shoot's response could have been positive phototropism.
c) Repeat the experiment in darkness and find out if the shoot still grows upwards.
4 a) **Y**.
b) Unlike leaf **X**, leaf **Y** has not undergone abscission and its lateral bud has remained dormant.

Experimental design answer

Living organism under investigation
Rose plant.

Prediction
A greater amount of root formation will occur in solutions of higher auxin concentration than in solutions of lower auxin concentration.

Experimental procedure
1 Make up a series of auxin solutions such as 10^{-6}, 10^{-5}, 10^{-4}, 10^{-3}, 10^{-2}, 10^{-1}, 1 and 10 parts per million (ppm).
2 Include pure water as a control.
3 Place the ends of 3 rose cuttings in each liquid in a container covered with light-proof material.

4 After a few weeks measure the length of all of the roots formed by each cutting and calculate an average for each condition.
5 Pool results with other groups.
6 Calculate the percentage stimulation of root formation for each condition using the formula

$$\frac{\text{average change in length relative to control}}{\text{average increase in length of control}} \times \frac{100}{1}$$

7 Compare the results and find out if they match the prediction and therefore support the hypothesis (or not).

28 Hormonal influences on growth – part 2

1 a) (i) It makes stem length increase.
(ii) Length.
b) (i) GA is produced in the embryo and exerts its effect in the aleurone layer where it induces the production of α-amylase.
(ii) Inaccurate cutting of the grain results in a portion receiving both embryo and endosperm tissue enabling it to make the enzyme.
2 a) Selective weedkiller (herbicide).
b) Unlike the narrow-leaved grass, the broad-leaved weed absorbs much of the chemical. This synthetic auxin stimulates the weed's rate of growth to such an extent that it dies of starvation.

3

role	G only	A only	G and A
induction of α-amylase production in cereal grains	✓		
stimulation of cell division and elongation			✓
reversal of genetic dwarfism	✓		
stimulation of adventitious root formation		✓	
prevention of leaf abscission		✓	
promotion of phototropic growth movements		✓	

Data interpretation answers

a) See figure A28.1.
b) (i) X.
(ii) The average total length of the side buds and shoots per plant remained unchanged until day 8.

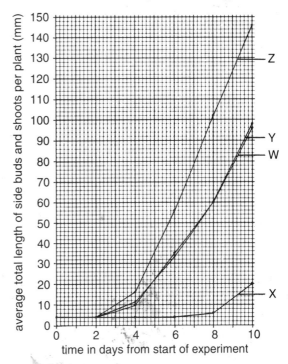

Figure A28.1

c) It acts as a control to check that plain lanolin has no effect on the plant.
d) (i) Gibberellin.
(ii) Drawing such a conclusion is justified because the average length of the side buds and shoots for group **Z** (with gibberellin) is even greater than control group **W** (untreated).
(iii)

$$\frac{\text{average change in length relative to control}}{\text{average increase in length of control}} \times \frac{100}{1}$$

$$= \frac{42}{56} \times \frac{100}{1} = 75\%$$

e) Perhaps the auxin supply in the lanolin was running out by day 10 and apical dominance was no longer being maintained.
f) The more organisms that are used, the more statistically valid the results.

29 Effects of chemicals on growth

1 a) Magnesium.
b) It is needed for chlorophyll formation.
c) There is a small supply of essential elements in the seed but it runs out after a few weeks.
d) A version of the experiment using complete medium.

2 a) 1500 g
   b) Liver, eggs, spinach.
3 a) X = 300; Y = 200.
   b) Some would be used for bone formation and therefore less would pass out in body wastes.
4 a) D.
   b) In Asia their skin had received plenty of ultraviolet radiation forming vitamin D.
   c) The child would be bow-legged.
   d) Cod liver oil.
   e) It promotes the absorption of calcium and phosphate from the intestine and their uptake by bone.

30 Effect of light on growth

1 a) (i) The leaves of an etiolated plant are small and yellow whereas those of a normal plant are large and green.
   (ii) The stem internodes of an etiolated plant are elongated and weak whereas those of a normal plant are short and strong.
   b) An etiolated bean seedling shows an increase in cell number but a decrease in dry mass. Whether this can be regarded as real growth is debatable. It depends upon which definition of growth is favoured.
   c) Death. Eventually it would run out of food.
   d) (i) The existing shoots will probably turn yellowish and some leaves may die but the leaves and stems will not change in structure.
   (ii) The new shoots will show etiolation with respect to both leaves and internodes.
2 See figure A30.1.
3 a) Leaf.
   b) They should be given short nights (i.e. less than the critical 9 hours of darkness) up until a few weeks before Christmas, and then be given the correct length of darkness to make them flower.
4 Arctic and temperate regions in summer. The photoperiod is long.
5 a) Wavelength of light.
   b) Intensity.
   c) (i) 500.
   (ii) 650+.
   d) It absorbs blue, green and yellow light but gives some of the violet and all of the red back out again making it appear purple (violet + red) to the naked eye.
   e) Pituitary.
   f) Enlargement of testes and onset of mating behaviour.
6 a) Length of photoperiod.
   b) Long.
   c) High.
   d) Breeding would begin even though it was the wrong time of the year.
   e) Daylength is absolutely predictable since its pattern of change is repeated in exactly the same way year after year. Temperature is unpredictable since 'freak' conditions such as cool summers and mild winters occur.
7 The bright light stimulated each patient's pineal gland. In most cases less melatonin was secreted and the patients lost their feeling of winter gloom. They felt as if it was summertime and their depression lifted. It returned when the light treatment stopped and their melatonin production increased again.

Data interpretation answers

a) (i) A = short day (11 hours of darkness)

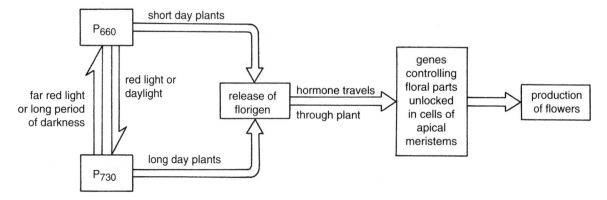

Figure A30.1

B = long day (12 hours of light)
C = short day (12 hours of darkness)
D = day neutral.
(ii) 12 hours of light and 12 hours of darkness.
b) C. Treatments **V** and **W** show that the plant used for the further series of experiments needs 12 hours of darkness to flower but fails to do so with less. This matches plant **C**.
c) P_{660} is needed by short day plants to flower. Treatment **X** applied red light which converted essential P_{660} to P_{730}. Treatment **Y** did the same but then applied far red light which reversed the process converting P_{730} back to essential P_{660}.
d) No flowering would occur since the overall effect would be the conversion of P_{660} to P_{730}.
e) The plants could be given artificially long days (i.e. 12 or more hours of light). The plants could be given short bursts of red light during the hours of darkness.

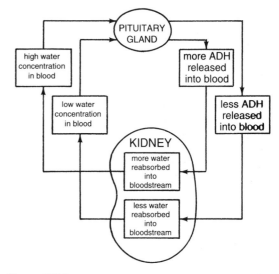

Figure A31.1

31 Physiological homeostasis

1 a) Physiological homeostasis is the maintenance of the body's internal environment within tolerable limits despite changes in the external environment.
b) It is a corrective mechanism. Receptors detect a deviation from the norm of a factor affecting the body's internal environment and send messages to effectors. These bring about responses which negate the deviation and return the system to its set point.
c) It is of survival value because it keeps the body's internal environment at an optimum state regardless of the external environment.
2 a) See figure A31.1.
b) (i) Antidiuretic hormone. It makes them more permeable to water.
(ii) The higher the amount of ADH in the bloodstream the lower the volume of, and the higher the concentration of, the urine formed. This is because ADH promotes reabsorption of water from the kidney tubules into the bloodstream.
c) Hypothalamus.
d) If water is drunk, it will pass from the gut into the bloodstream and return the blood to its normal water concentration.
3 a) (i) 07.00–08.00.
(ii) Physiological homeostasis.
b) (i) 08.00.
(ii) They both increased.
(iii) It took a little time for the blood containing extra glucose to arrive at the pancreas and for

the pancreas to respond and make more insulin.
c) As the concentration of insulin increased, the concentration of fatty acids (the breakdown products of fat) decreased. This suggests that insulin suppressed the breakdown of fat to fatty acids.
4 a) and **b)**

hormone	endocrine gland from which hormone originates	letter(s) indicating effect(s) of hormone
adrenaline	adrenal	**C** and **E**
insulin	pancreas	**A** and **D**
ADH	pituitary	**B**
glucagon	pancreas	**C**

5 a) Osmoregulation involves the maintenance of the correct water concentration of the blood. This is effected homeostatistically by the required amount of water (but not urea) being reabsorbed from glomerular filtrate. Since the urine formed contains all of the original urea but less water, the concentration of urea in urine is higher than in glomerular filtrate.
b) (i) 2.
(ii) 70.
c) No urea is reabsorbed but some sodium is reabsorbed.
d) (i) Glucose would have been present in the urine.
(ii) Insulin and a controlled diet.
6 Eventually negative feedback breaks down

when the corrective responses (e.g. shivering and vasoconstriction) fail to return the body to its set point (normal body temperature).

Data interpretation answers

a) Blood plasma.

b) (i) Adrenal gland.
(ii) Kidney.

c) Hormonal.

d) Negative feedback control involving reduced secretion of aldosterone (and reduced reabsorption of salt).

e) See figure A31.2.

f) (i) This promotes the absorption of extra salt needed to replace losses.
(ii) Anti-diuretic hormone (ADH).

g) The person would continue to lose salt in sweat and so more and more aldosterone would be secreted to promote maximum reabsorption of salt from glomerular filtrate. However the body's salt content would still continue to decrease and eventually the homeostatic mechanism would break down. This would lead to death.

32 Population dynamics – part 1

1 a) 55/m².
  b) 35 pairs.
  c) To give a balanced overall picture which is unaffected by any unusual results.
  d) Once the *population* size of a species of *plant* or animal has reached the *carrying* capacity of the *environment*, it remains relatively *stable* despite short-term *oscillations* in number.

2 a) (i) A density-dependent factor is one which only affects the population once it has grown to a certain density.
(ii) A density-independent factor is one which affects the growth of a population regardless of the population's density.
  b) Density-dependent = increased predation and shortage of food; density-independent = a thunderstorm and intense drought.

3 a) Increase in daylength allows more food to be made by photosynthesis leading to increase in cell numbers. Rise in temperature of water increases rate of enzyme action in cells promoting growth and multiplication.
  b) See figure A32.1.

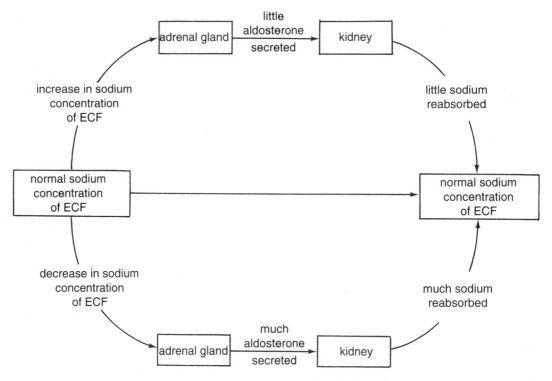

Figure A31.2

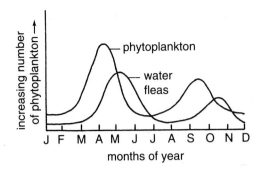

Figure A32.1

4 a) Number of predators, amount of available food and incidence of disease.
b) Homeostasis.
5 a) (i) 35.
(ii) 33.
b) 100.
c) (i) Lack of available prey to eat.
(ii) Fewer predators were eating them.
6 a) Density-dependent. It only affects the dense populations in the burrows and not the solitary individuals above ground.
b) (i) See figure A32.2.

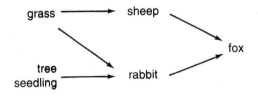

Figure A32.2

(ii) There will be more grass for the sheep to eat but the farm animals will suffer more attacks by foxes.

Experimental design answer

Living organism under investigation
Springtail.

Prediction
Most of the springtails in an area of soil sprayed with 1% NABBEM will be killed.

Experimental procedure
1 Choose a stretch of soil known to contain springtails.
2 Select at random a few sample sites each of uniform area.

3 Enclose each with a wooden frame and spray the soil with a standard volume of 1% NABBEM.
4 On the same stretch of soil repeat the procedure for new sites using water instead of the pesticide.
5 Using a soil auger, remove several soil samples of known volume from each of the treated sites.
6 Separate the springtails from each sample using Tullgren funnels.
7 Calculate the average population density (springtails/cm^3) for soil sprayed with NABBEM and soil sprayed with water.
8 Compare the results and find out if the prediction is correct and the hypothesis is supported (or not). If it is supported, gardeners should be advised to stop using NABBEM until a safe concentration (if this exists) is established by further experiments.

Data interpretation answers

a) (i) Predator = broken line, prey = solid line.
(ii) The solid line shows a trend of population fluctuations which are mirrored by the broken line after a time lag. This is typical of a predator-prey graph. The solid line represents the numerically larger group which would be true of a prey organism compared with its predator.
b) (i) 10.
(ii) When prey numbers increase, the predators have access to abundant food supplies and after a time they also increase in number. The reverse also applies.
(iii) If the unpalatable grass is the real cause of the rabbits' periodic decline in numbers then tracking is a more accurate way of describing the predators' numbers, since they passively follow rather than actively cause the preys' decline.
c) (i) Limpet and chiton.
(ii) In the absence of starfish, barnacles increased in number and won out in the competition for the algae.
(iii) It keeps the number of barnacles down.
d) Starfish is a 'generalist' and lynx is a 'specialist'.

33 Population dynamics – part 2

1 a) It means keeping populations under close surveillance with respect to their numbers and factors which affect them.
b) It is necessary so that information is available about food species, endangered species and indicator species (allowing human intervention where appropriate).

2 Taking the smaller younger fish (in addition to the adults) would mean removing fish at a rate that exceeds their maximum rate of reproduction. In the long term this would lead to extinction of the species.

3 a) It means a temporary prohibition being placed on the activity.
b) It is essential in order to save those species of whale that are already threatened with extinction.

4 a) It means a species which shows the state of the health of its environment.
b) (i) The more animal material (especially fish) that the bird eats, the higher the concentration of pesticide residue in its muscle tissues.
(ii) Animals such as fish are situated further along the food chain than plants or invertebrates and therefore eat food which has already concentrated pesticide at several links in the chain.
c) The sea. The sea dilutes the concentration of pesticide arriving in rivers before it enters the marine food chain.

5 a) They support it in the belief that it will help to prevent the elephant from becoming extinct.
b) If there is no demand for ivory goods then there is no point in killing elephants for their tusks.
c) Zimbabwe has carefully managed its population of elephants and therefore feels that it is being unfairly punished for the poor practices that take place in other African countries.
d) 70%.
e) It means the planned harvesting of a wildlife species.
f) (i) Work and free meat.
(ii) Destruction of trees and grassland is prevented.
g) Perhaps an attempt could be made to introduce Zimbabwe's good management practices to other countries. If this proved to be a success, some countries might eventually decide to drop the moratorium. (Or . . . ?)

Data interpretation answers

a) An inverse relationship exists. As the bacterial numbers increase they consume more and more of the dissolved oxygen which therefore decreases.
b) (i) It is absent from the region of the river upstream from the sewage inflow.
(ii) It increases because essential factors for growth (such as food, oxygen, water etc.) are available.

(iii) It could have run out of food (or oxygen).
(iv) Density-dependent.
c) (i) Lack of light for photosynthesis (owing to the dirty, cloudy state of the water).
(ii) Density-independent. The sudden reduction in light intensity in the murky water following the sewage inflow affects the population of algae regardless of whether the population is dense or sparse.
d) In the nitrogen cycle the sequence of events is:

ammonium compounds $\xrightarrow{\text{bacteria}}$ nitrites $\xrightarrow{\text{bacteria}}$ nitrates

e) Nitrate ions are needed by plants to make protein. A rich supply of nitrate therefore promotes plant growth.
f) Heavy rain.

34 Population dynamics – part 3

1 a) Succession means a change involving the regular progression from a pioneer community of plants to a climax community.
b) Primary succession occurs during the colonisation of a barren area. It takes a considerable length of time. Secondary succession occurs during the colonisation of an area which has been previously occupied. It takes a shorter time.

2 a) A climax community is the final product of succession. It is self-perpetuating and not replaced by another community.
b) Climate and soil type.
c) Tundra, broad leaf forest and mediterranean scrub.

3 a) 6.
b) (i) 8.
(ii) G, H, I and J.
(iii) 3.
(iv) The soil was still not sufficiently fertile.

4 See figure A34.1.

Data interpretation answers

a) Primary succession, since this is colonisation of a barren area which has not been previously inhabited.
b) (i) It forms hummocks held together by the plant's roots.
(ii) Marram grass.
(iii) Its extensive roots stabilise more sand and its dead remains add humus to the sand.
(iv) Sea holly, fescue grass and heather are able to succeed marram grass because the latter has formed a thin layer of soil.

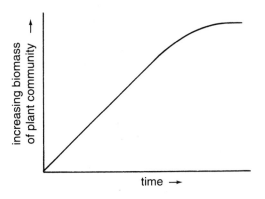

Figure A34.1

c) (i) **D**.
(ii) Yes. Unlike regions further inland, the soil at **B** continues to be affected by loose sand blown up from the beach. Marram grass can be described as an edaphic climax since succession fails to proceed beyond it on this atypical soil.
(iii) **A**.

35 Maintaining a water balance – animals

1 a) The left bar in the graph shows that the fresh water fish is hypertonic to the surrounding water (which has a much lower concentration of salt). As a result the fish tends to gain unwanted water by osmosis.
The right bar in the graph shows that the salt water fish is hypotonic to the surrounding water (which has a much higher concentration of salt). As a result the fish tends to lose water by osmosis.

b)

	fresh water bony fish	salt water bony fish
relative size of glomeruli	large	small
relative number of glomeruli	many	few
relative filtration rate of blood	high	low
relative volume of urine	large	small
relative concentration of urine	dilute	concentrated
direction in which salts are actively transported by chloride secretory cells	into body	out of body

c) Diffusion is the movement of molecules along a concentration gradient from high to low, whereas active transport is the movement of molecules against a concentration gradient from low to high. Diffusion does not require energy; active transport does require energy.

2 a) (i) In food and metabolic water.
(ii) In urine and exhaled air.
b) Physiological = production of high level of anti-diuretic hormone to promote maximum reabsorption of water; lack of sweating. Behavioural = fairly inactive during the hot day; active during the cool night.

3 a) (i) Shaded.
(ii) Unshaded.
b) On moving from fresh to salt water, the filtration rate of the eel's kidney decreases, a reduced volume of urine is made and the chloride secretory cells actively transport salt out of the body. Each of these is reversed when the fish moves from salt to fresh water.
c) The adult eel migrates to salt water to spawn, whereas the adult salmon migrates to fresh water to spawn.
d) Fish able to adapt and migrate are able to exploit a wider range of habitat.

4 a) **B**.
b) (i) **A**.
(ii) 0.5.
c) (i) **A**.
(ii) Oxygen is used during aerobic respiration to generate the energy needed to actively transport molecules against a concentration gradient.
d)

habitat	species of crab	reason
river estuary	A	It is able to exert some degree of osmoregulation. This suits it to life in a habitat where the salt concentration of the water varies with the tides and the river flow.
deep sea water	B	Although it is unable to osmoregulate, it is well suited to its habitat because the salt concentration of the sea water remains constant at 3.5%

36 Maintaining a water balance – plants

1 a) (i) It enters a root hair by osmosis and moves across the root from cell to cell along a concentration gradient. It passes up through the xylem and across the leaf from cell to cell along a concentration gradient by osmosis. It diffuses out of a cell and becomes water vapour in a

moist air space. Finally it passes out through a stoma.

(ii) It is taken up by a root hair employing active transport. It moves across the root from cell to cell via tiny cytoplasmic connections and is actively transported into a xylem vessel. It is passively carried up the xylem and actively transported into a green leaf cell where it is used to make chlorophyll.

b) Transpiration stream.

2 **a)** To prevent air entering the xylem vessels and forming air locks.

b) The rubber tubing must be a tight fit to prevent entry of air and leakage of liquid contents.

c) Cohesion-tension theory. The fact that the ascending water molecules are able to pull up a column of mercury (a very dense liquid) without the system breaking down, shows that the water molecules must be cohering strongly together.

3 **a)** It decreases during the day and increases at night.

b) During periods of maximum transpiration, the columns of water molecules in the xylem vessels are pulled thin by the tension from above. At the same time they adhere to the insides of the xylem vessel walls pulling them inwards. This has an overall effect of reducing the tree's diameter.

4 **a)** Under such circumstances, all leaf cells including guard cells continue to lose turgor until the stomata close. Water loss is therefore reduced until conditions improve.

b) The system could break down and the plant could die from lack of water following a long period of drought. This is because a little water continues to be lost directly through the leaf surfaces even when the stomata are closed.

5 **a)**

	apparatus X	apparatus Y
initial mass (g)	283.80	455.72
mass after 2 days (g)	259.32	399.72
mass of water lost (g)	24.48	56.00
volume of water lost (cm^3)	24.48	56.00
volume of water required to restore initial level (cm^3)	25.00	56.00

b) 0.51 g.

c) More water is taken in by the plant than given back out again because the plant retains a little

for photosynthesis and maintenance of turgor.

d) (i) Rate of water loss will decrease in both **X** and **Y**.

(ii) Rate of water loss will come to a halt (or very nearly) in **X** but it will continue as normal in **Y**. In **X** the stomata will close in darkness and prevent transpiration. In **Y** loss of water by evaporation will not be affected by darkness.

e) (i) Weight atmometer.

(ii) It acts as a type of control by indicating when **X** is acting as a free evaporator and when it is affected by physiological factors such as stomatal closure.

6 **a)** See figure A36.1.

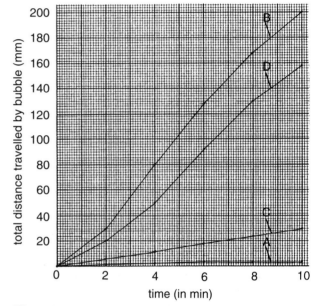

Figure A36.1

b) **B**. None of its stomata were blocked.

c) **A**.

d) The fact that the bubble moved a short distance in **A**.

e) Rate of water uptake by the plant.

f) Lower. A comparison of **C** and **D** shows that **D** (lower surface free of vaseline) lost more water than **C** (upper surface free of vaseline).

7 **a)** Each leaf is reduced to a needle which has a thick epidermis and sunken stomata.

b) Needles reduce size of overall transpiring surface (compared with broad-leaved tree). Thick epidermis reduces amount of water lost directly through the leaf surface. Each sunken stoma is separated from the outside air by a pit of moist air which helps to reduce water loss.

8 **a)** Xylem is not needed for water transport since the plant's cells have easy access to the

surrounding water. Little xylem is needed for support since this is provided by the surrounding water.

b) They give the plant buoyancy keeping its leaves near the water surface for light. They store oxygen (made during photosynthesis) for later use.

Experimental design answer

Living organism under investigation
Iris.

Prediction
If stomatal numbers are equal on both leaf surfaces then the rate of water loss from leaves with their upper surfaces coated with vaseline will be equal to that of leaves with their lower surfaces coated with vaseline.

Experimental procedure
1 Separate a young non-flowering iris shoot from its rhizome by cutting it under water.
2 Connect the shoot to tightly-fitting rubber or plastic tubing, fitted to capillary tubing to make a bubble potometer which is completely free of air locks.
3 Coat all of the shoot's upper leaf surfaces with vaseline.
4 Repeat the procedure using a second shoot equal in size to the first but coat its lower leaf surfaces with vaseline.
5 Allow time for the plants to acclimatise and then introduce a bubble into each potometer.
6 Measure the distance moved by the bubble per unit time for both potometers kept in equal environmental conditions.
7 Pool class results and calculate averages.
8 Compare the two sets of results which indicate the rate at which water was lost by transpiration from each type of leaf surface. If they are equal then they agree with the prediction and the hypothesis is valid.

Data interpretation answers

a) (i) Increase in temperature reduces time taken.
   (ii) Increase in wind speed reduces time taken.
   (iii) Darkness increases time taken.
b) There are two variable factors involved at the same time.
c) 9.
d) 4 and 8.
e) 5 and 6.
f) (i) Air humidity.
   (ii) As air humidity increases, transpiration rate decreases.
   (iii) During transpiration, water vapour

molecules diffuse along a concentration gradient from high concentration inside the leaf's moist air spaces to low concentration in the air outside. As the humidity of the air ourside increases, the concentration of water vapour molecules increases and the concentration gradient becomes less steep resulting in decreased rate of water loss by transpiration.

37 Obtaining food – animals

1 a) It means to go searching for food.
   b) It must find food that gives it more energy than the amount it expends on the search or it will suffer a net loss and eventually die.
2 a) (i) A repeat of the experiment with a chemical-free trail.
   (ii) Otherwise it could be argued that the ants were simply following the pattern of disturbed soil and not the chemical.
   b) Ants locate food by following a trail containing chemical from a previous ant's body.
   c) A repeat of the experiment without food at the end of the trail.
   d) As long as food is present, the ants passing along the trail keep adding more chemical which is picked up by the sense organs of later ants and so on.
   e) No chemical is added to the trail when the food runs out so the trail soon lacks scent and fails to attract any ants.

3

feature of experimental design	reason
same number of animals of same species used and identical environmental conditions maintained in each dish	to ensure that only one variable factor is under investigation
several animals used in each dish	to ensure that the results are not based on only one animal whose behaviour might not be typical of the species in general
hungry animals used	to ensure that the animals are motivated and ready to search for food
glass bead similar in size to the piece of liver used in the control	to ensure that the animals are not simply attracted to any physical object
experiment repeated many times	to ensure that the results are typical and valid

4 a) Number of bats; species of bat; size of

environmental chamber; temperature of
environmental chamber.
b) Sense of hearing (one group lack it).
c) Bats with their ears plugged will fail to catch
prey and will tend to collide with the wires
because they will be unable to pick up echoes
from objects in their path.
5 a) Ears, hackles and tail raised; teeth bared.
    b) Ears, hackles and tail lowered; teeth covered.
    c) Dominance hierarchy.
    d) It keeps real aggression down to a minimum
and guarantees experienced leadership.
6 a) A territory is a small area (usually containing
food) which an animal establishes as its own
private domain. Territoriality is a form of
intraspecific competition between animals for
available territories.
    b) An example is a robin defending its territory
using visual and auditory signals to warn off
rivals.
    c) It results in the animals being spaced out in
relation to the available food supply and being
provided with safe places to breed.
7 a) (i) 5 joules/second.
    (ii) 6 joules/second.
    (iii) 5 joules/second.
    b) (ii)
    c) The predator may have to expend too much
energy subduing the prey.

Data interpretation answers

a) **A** = dry and cold, **B** = wet and hot.
b) Wet and cold. Since this set of conditions
provides each species with one (but not both) of
the conditions that it favours, the two species
begin roughly equal and fierce competition
follows.
c) It is supported in that under both sets of
extreme conditions, one species ousts the other.
d) Both species would survive with neither
gaining the upper hand for a long enough time
to oust the other.
e) Graph **1** shows that when narrow tubes are
present, the number of species **C** increases;
graph **2** shows that when wide tubes are
present, the number of species **C** drops to zero.
It is possible that **A** eats **C** but that **C** can escape
and multiply by inhabiting glass tubes which
are too narrow for predator **A** to enter.
f) Interspecific.

38 Obtaining food – plants

1 Mobility is the ability to move the whole body

from place to place. Sessility is the state of
remaining fixed to one position and being
unable to move from place to place.
2 a) Water and soil nutrients.
    b) Interspecific competition is less intense
because, unlike intraspecific competition, it
involves different species which do not have
exactly the same requirements.
3 a) See figure A38.1.

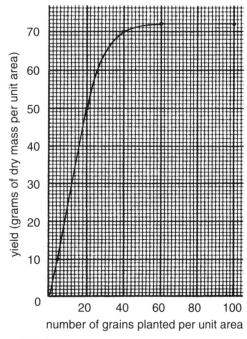

Figure A38.1

    b) (i) As the number of grains increases so does
the yield.
    (ii) Eventually the grains will be so numerous
on the unit of soil that they will be competing
with one another for any resources that are in
limited supply.
    c) A parasite could spread more easily from
plant to plant.
    d) **A** (since there would be plenty of room for
weeds).
    e) 60.
4 a) **B**.
    b) The leaf (or flower). Perhaps the inhibitor is
washed off the leaf (or flower) surfaces as water
flows over them from above. A comparison of
pots **B** and **C** suggests this to be the case.
    c) To show that the physical effect of water
landing on flax plants from above is not the
factor that inhibits their growth.
    d) More plants should be used in each pot. The
experiment should be repeated several times to

see if the results can be successfully repeated.

5 X had 20; Y had 11. Cropping maintains species diversity by keeping the sturdier plants in check and preventing them from choking out the more delicate ones.

6 **a)** Compensation point is the particular light intensity at which a plant's rate of photosynthesis exactly equals its rate of respiration.

Compensation period is the time taken for a plant which has been in darkness to reach its compensation point.

b) True. The plant must get beyond this point regularly so that photosynthesis exceeds respiration and a store of food can be made for use during the night and on dull days.

c) (i) 1 hour 40 minutes approximately.

(ii) 6 am and 10 pm.

(iii) 16 hours.

(iv) It would have been less.

Experimental design answer

Living organisms under investigation
Related species of sun plant and shade plant.

Prediction
A chromatogram of photosynthetic pigments from the shade plant will possess much larger spots of the yellow carotenoid pigments than that of the sun plant.

Experimental procedure
See Experimental design, chapter 5.

39 Coping with dangers – animals

1 Habituation. It prevents the animal from wasting energy on many needless repeats of its escape response.

2 **a)** To ensure that the results are not based on the behaviour of only a few animals which might not be typical of the whole species.

b) By tapping the side of the tank.

c) It withdraws its head into its tube.

d) (i) In both cases, after several trials, the number of worms showing the escape response had dropped to zero.

(ii) At trial 20 when use of the second type of stimulus (moving shadow) began, the percentage number of worms showing the escape response was as high or higher than it had been at trial 1 when use of the first stimulus (mechanical shock) began. However all of the worms had become habituated to the first type of stimulus by trial 20, so the second type of

stimulus must have been acting independently of the first.

e) (i) 25.

(ii) Since it is short-lived, the animal soon begins to show its escape response again which saves it from real danger.

3 **a)** Learning curve.

b) He was still learning how to do the job.

c) Between 3 and 4. He had learned how to do the job and had remembered what to do from one day to the next thereby improving his performance.

d) He had reached his best level of performance.

4 **a)** Counter shading.

b) Viewed from above, the bird's dark colour tends to blend with the dark land mass below. Viewed from underneath, the bird's light colour tends to blend with the bright sky above.

c) Individual.

5 **a)** It is possible that this adaptation works by confusing the predator. Any moment of delay on the part of the lion will give the herd of zebra a chance to escape.

b) Social.

6 Camouflage relies on the caterpillars being widely scattered. If they cluster together they become conspicuous and are easily picked off by predators.

Bright warning colours are more effective when the caterpillars are clustered together because the message becomes even more obvious to potential predators.

Experimental design answer

Living organism under investigation
Larvae of common gnat (*Culex*).

Prediction
Freshly hatched gnat larvae will at first show their escape response to water disturbances caused by bubbles from an aerator, but will eventually fail to do so if the bubbling is maintained continuously.

Experimental procedure

1 Obtain a supply (e.g. 20) of freshly hatched gnat larvae that have not previously been exposed to water disturbances caused by bubbles from an aerator.

2 Label two tanks of pond water A and B and keep all environmental factors affecting them equal.

3 Introduce an aerator to each tank but keep it switched off at present.

4 Add 10 gnat larvae to each tank and give them time to acclimatise.

5 Turn on the aerator in Tank A and note the number of larvae that show the escape response. Leave the aerator on for the duration of the experiment.

6 Turn on the aerator in tank B (the control) for 10 seconds only and note the number of larvae that show the escape response.

7 After 2 minutes note the number of larvae showing the escape response in tank A, then repeat step 6.

8 Repeat this procedure at 2-minute intervals for 30 minutes.

9 Pool results with other groups and calculate averages.

10 Find out if the results match the prediction and lend support to the hypothesis (or not).

Data interpretation answers

a) Successful camouflage works best when the organism does not move.

b) Ability to sting; possession of poison.

c) The mimic. The predator is deceived into avoiding palatable prey. The model's security is threatened if the predator takes a chance after finding a mimic palatable.

d) By imitating the sound made by the stinging/poisonous model, the harmless mimic scares off the predator.

e) They share any losses suffered while the predator is learning the warning signs.

f) Batesian.

g) Mullerian. Since the warning colours are ineffective in darkness, an odour warns potential predators instead.

40 Coping with dangers – plants

1 a) It is the production and release of poisonous hydrogen cyanide by a plant.

b) The presence of poison deters most herbivorous animals from eating the plant.

c) The hairy potato plant produces a chemical which is so similar to the alarm scent of aphids that it keeps them away.

2 a) See figure A40.1.

b) The younger the leaf, the higher its cyanide content.

c) The presence of a higher concentration of cyanide in young leaves protects them against hungry animals at a time when the leaves are delicate and especially vulnerable to attack.

3 They have low growing points and underground rhizomes (organs of asexual reproduction) which enable them to recover and

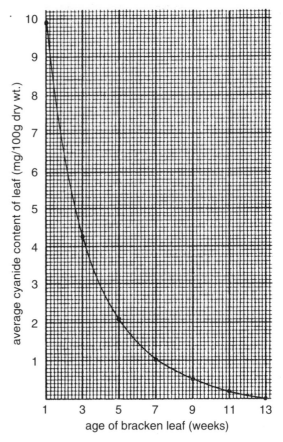

Figure A40.1

grow again even after most of their leaves have been eaten.

4 a) An inverse relationship. The number of spines decreases as the distance of the leaf from the ground increases.

b) It is of survival value to have protective spines on leaves that are within reach of browsing herbivores. However spines are of no value to leaves further up the tree and safely out of reach.

c) Stinging hairs found on nettle leaves. On contact with the victim, the tip of the hair breaks and injects an irritant which warns the victim off.

Data interpretation answers

a) Acyanogenic.

b) GGEE, GGEe, GgEE and GgEe.

c) (i) glycoside $\xrightarrow{\text{enzyme}}$ hydrogen cyanide.

(ii) This does not happen because formation of hydrogen cyanide only occurs when the leaf is damaged by an external agent.

d) (i) The different species of invertebrate eat the acyanogenic plants and reject the cyanogenic ones.

(ii) 3 and 6.

(iii) 70%.

(iv) They are able to survive by repelling most herbivorous invertebrates.

e) (i) It is inverse. As altitude increases, capability of cyanogenesis decreases.

(ii) 100%.

(iii) Since they will be affected by intense frost for many months of the year, it is best that they do not make hydrogen cyanide or they would poison themselves to death.

Notes